Writing and Presenting Scientific Papers

2nd Edition

Writing and Presenting Scientific Papers
2nd Edition

Birgitta Malmfors
Phil Garnsworthy
Michael Grossman

NOTTINGHAM
University Press

First Published by Nottingham University Press
This reissued original edition published 2023 by 5m Books Ltd
www.5mbooks.com

British Library Cataloguing in Publication Data
Writing and Presenting Scientific Papers, 2nd Edition:
I Malmfors, B., II Garnsworthy, P.C., III Grossman, M.

ISBN   9781789183214

Typeset by Nottingham University Press, Nottingham

EU GPSR Authorised Representative
LOGOS EUROPE, 9 rue Nicolas Poussin, 17000,
LA ROCHELLE, France
E-mail: Contact@logoseurope.eu

CONTENTS

THE AUTHORS

Birgitta Malmfors is an Associate Professor in Animal Breeding and Genetics at the Swedish University of Agricultural Sciences, Uppsala, Sweden. She has written many scientific papers, reports and articles. She has authored textbook chapters on animal breeding and a handbook on scientific writing and oral presentation, and also authored chapters on teaching methodologies published in a CD. She regularly makes presentations at international meetings. She has taught students successfully for 30 years, and now also trains university teachers in international courses. She has been given a distinguished award for teaching and communication skills.

Phil Garnsworthy is a Senior Lecturer in Animal Production at the University of Nottingham, UK. He has edited 18 books of conference proceedings and contributed chapters to several other books in the area of animal nutrition. He also has written many papers for scientific journals and articles for the popular press. He regularly makes oral and poster presentations at international scientific conferences and industry meetings. He has been teaching students for more than 20 years and is particularly interested in training students to communicate effectively.

Michael Grossman is a Professor in the Departments of Animal Sciences and Statistics at the University of Illinois at Urbana-Champaign, USA. He has written numerous scientific papers as well as popular science articles, and has given national and international scientific presentations. He has taught animal genetics and mathematical modelling for more than 30 years. He also has taught many workshop courses on techniques for writing and presenting a scientific paper to PhD students, both at the University of Illinois and at Wageningen University, The Netherlands, where he has received a prize from each university for his teaching skills.

Since 1998, the three authors have jointly presented a workshop on "Writing and Presenting Scientific Papers" in advance of the annual meeting of the European Association for Animal Production. This workshop continues to attract participants from many countries, who indicate that they benefit greatly from the experience.

REVIEWS OF THE FIRST EDITION

"This book should be required reading for all students whose degrees or professions will involve presenting scientific results in writing or orally. The topic is so important and the book so competent, it will quickly attract a grateful readership of both students, research scientists and teachers. It is well overdue and well done."

EAAP News & Livestock Production Science

"… a 'must read' for all scientists. Concise, practical information is provided that will be useful to both incoming graduate students and established scientists. Every person who wishes to communicate science effectively should have a copy of this book."

Ellen Bergfeld, Executive Director,
American Society of Animal Science

"…it is a useful book for both authors and reviewers. The information presented is clear, appropriate and concise. It is easy to read and much of the advice is applicable to non-scientific writing and presentation."

The Horticulturalist

"… this slim volume stands out as a classic of its genre … Not only is this book aimed at young scientists presenting their first results …but also to older members of the profession, wishing to improve their communication skills. In my opinion, it has the potential to do this very successfully … compulsory reading for all science undergraduates."

Annals of Botany

"… a highly valuable resource for science students, teachers, and researchers working in all disciplines."

Forest Science

"The book is very readable, is clearly presented, and makes humorous use of clip-art to highlight major points …The book presents clear and concise advice and instruction on topics that young scientists often find incredibly daunting"

Society for Experimental Biology

PREFACE TO THE SECOND EDITION

We were delighted that the first edition of this book was received positively. Many students and experienced scientists, including reviewers, told us that they found the book enjoyable and useful, with much good practical advice, and thus a valuable tool for improving writing and presentation skills.

The first edition was intended to be Accurate, Brief and Clear, yet broad enough to provide a comprehensive guide to most aspects of writing and presenting scientific papers. Feedback from readers, together with our further experiences of presenting workshops on science communication, however, suggested that the first edition could be expanded.

In the present edition of the book, new sections have been added on writing statistics and on writing research proposals. We have also added a section on book reviews. Other sections have been rearranged and expanded. In addition, we have taken the opportunity to revise the text to make it even clearer. Advances in technology have resulted in an increase in the use of electronic methods for preparation and submission of manuscripts, for handling references, for consulting electronic journals and for delivering conference presentations. In this edition, therefore, more emphasis has been placed on electronic media.

We dedicate this second edition to our families, with thanks for their patience and support during our work with the book. Thanks are due also to Sarah Keeling for typesetting, and to colleagues who reviewed early versions of new sections.

We hope that this book will help you to improve your writing and presentations. We emphasise again that our intention is not to dictate how to write and present. Our aim is to provide guidelines and suggestions, based on our collective experience.

Birgitta Malmfors
Phil Garnsworthy
Michael Grossman

November 2003

PREFACE TO THE FIRST EDITION

Think of any great scientists in history and consider what made them great. You will probably say that they made a new discovery, invented something or interpreted existing knowledge in a new way. How do we know they did this? They communicated their findings to others! Without communication, science would not have developed. Even today, scientists are judged by their ability to communicate their ideas and findings through written papers and conference presentations. You can perform the most elegant research in the world, but it has no value unless you tell somebody the results.

Preparing papers and presentations can take quite a long time, but usually not as long as the time spent doing the research. Publication and presentation should be seen as an integral part of the research process – it is the only way to produce something that people will remember. Scientific communication requires a special language, so that your message is clearly and accurately conveyed to the reader or listener; its effectiveness can be enhanced by a few simple techniques. Young scientists and students can be trained in these techniques, but experienced scientists can also benefit from critically examining the way they write and present papers.

This book is dedicated to Professor Rommert Politiek, The Netherlands, former Editor-in Chief of the Elsevier journal "Livestock Production Science". He identified the need for improvements in many papers submitted, as well as in oral and poster presentations at scientific meetings. He initiated, with great enthusiasm, a workshop for the European Association for Animal Production (EAAP), to train young scientists in writing and presentation skills. The first workshop took place at the annual meeting of EAAP in Warsaw (1998). The workshop was so successful that it was repeated in Zurich (1999) and again in The Hague (2000). The first two workshops were sponsored by EAAP and the third jointly by Elsevier and EAAP. We three authors of this book were encouraged by Professor Politiek to be the main tutors at these workshops. Having produced a booklet for the first workshop, we decided to gather our ideas together in the form of a book that would reach a wider audience.

Other books have been published on various aspects of writing and presenting scientific papers, some covering just one facet, others going into much more depth; a number of these books are included in our listing of literature for further reading. Our book is intended to be Accurate and Audience-adapted, Brief and Clear (the communication ABC). It covers the whole "chain" of science communication, including literature searching, writing for journals, conferences and the popular press, presenting papers orally or as posters, and training students in communication.

The book is aimed at scientists of all ages, university students and university teachers – anybody involved in communicating research results, or in training students. We particularly hope that young scientists will find the book useful and that it will encourage them in their communications. Although the EAAP workshops were designed for Animal Scientists, the information has been adapted to be useful for scientists and students from all disciplines.

Producing this book has been an interesting and valuable experience for us. We have learnt techniques from each other and we have broadened our perspectives. Although we all speak "English" (native and non-native), our origins are Sweden, UK and USA. This mix of writing cultures has emphasised to us that there is usually no "correct" way to write or present. For the convenience of our UK publisher, the book is written mostly in "British" English, but we have always tried to remember the need to keep the message clear for an international audience.

All drafts were produced directly on the computer, using ordinary software for word processing, graphics and presentation. Many of the illustrations were produced using Clip Art from Microsoft under the terms of their end-user agreement, which allows licensed users to use clip art in any publication, providing the publication itself is not a collection of clip art. We found electronic mail invaluable for the exchange of documents and ideas, which allowed us to work from three different countries.

We also received very valuable help while working on the manuscript; special thanks are addressed to Professor Jan Philipsson, Sweden, for reading drafts and providing useful suggestions, to Sarah Keeling, UK, for typesetting the manuscript, and to Nancy Boston, UK, for reading the final proofs.

Do not expect to find "rules" in this book; it's the journals, publishers and conference organisers that set rules for you to follow. Each of us writes and presents in different styles, so we do not have any definitive answers. What we do provide, however, are guidelines for good practice, based on personal experience and observation. We hope that you will find these guidelines useful in your scientific communications.

Birgitta Malmfors
Phil Garnsworthy
Michael Grossman

July 2000

1

COMMUNICATING SCIENCE

The aim of research is to contribute to knowledge, so that every new result adds to the previous state of knowledge, forming a basis for new thinking and interpretation, new implications, identification of needs for further research, etc. Research results, however, do not contribute to knowledge and development unless they are communicated effectively. Effective communication of science, to scientists and to other audiences, is an important component of the research process.

Communication is needed, not only to spread research results, but also to articulate results. Writing or talking about your research helps to clarify your thoughts and to put your research into a deeper and wider context. Therefore, start writing and talking about your research early in the research process.

Research results need to be communicated to contribute to new knowledge

Scientific communication occurs in many forms, such as papers in scientific journals, reports, conference papers and abstracts, theses and dissertations, review papers, proposals, popular science and newspaper articles, computer-mediated information, oral and poster presentations, interviews and discussions. Various forms of communication may differ, e.g. in purpose and audience addressed, but they also have a great deal in common – they communicate science. A good knowledge of how to handle various forms of scientific communication is needed in many professions, not only for scientists. It is essential, therefore, to include training in scientific communication at every level of education.

THE ABC OF SCIENCE COMMUNICATION

Communicating science usually means communicating new knowledge or summarising the present state of knowledge. It is important, therefore, that what is written or said is unambiguous, so that the audience understands the message.

The ABC of science communication is that it should be:

- Accurate and Audience-adapted
- Brief
- Clear

Science is international. This means that many of those who read a scientific paper or listen to a scientific presentation will be doing so in a foreign language. This further emphasises the need for clarity and for the presentation to be logical, consistent and coherent. Communication is a two-way process. Information cannot merely be delivered - it must be received and understood as well. The message delivered may be accurate, brief and clear, but may still not be received and understood. This can happen if what you write or say does not relate to the frames of reference of the audience. Adapting to the audience, therefore, is important.

A basis for the scientific process is to formulate a hypothesis, which means that you pose a question and a hypothetical answer. Questions and answers are the basis for communication as well. For effective communication you cannot just think of your own topic and the message you want to deliver. You must also consider what questions your audience might have. Some components of effective communication are indicated in Figure 1:1.

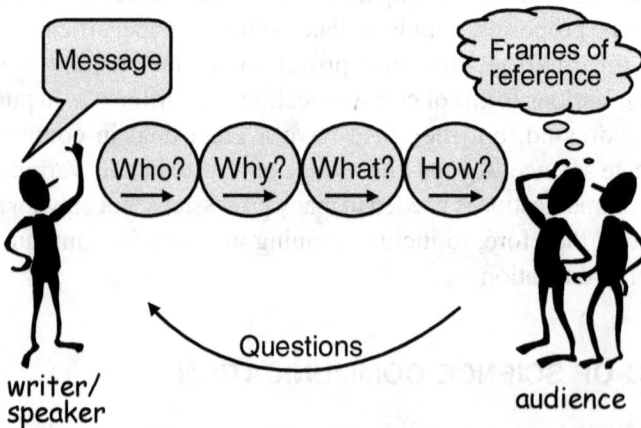

Figure 1:1. Some components of effective communication.

When preparing to write a paper or a report, or to make an oral or poster presentation, start by asking yourself the questions: Who? - Why? - What? - How?

- **Who** are you addressing: scientists who are specialists in your field of research, a wider group of scientists, fellow students, or public audiences?

- **Why** is your message important? Why are you communicating it? Presumably you are not doing it just for credits, but to add to the pool of knowledge, to teach, to inform, to persuade or push for development.

- **What** are your main findings or "take-home" messages? What are you going to present - new research results or a review of a topic? What prior knowledge, expectations and questions might your audience have? What technical language do they understand?

- **How** can you best deliver your message and satisfy the audience's needs? How will the audience use its new knowledge?

The order of the words, who-why-what-how, and their relative importance may vary in different forms of scientific communication. This is also true for the emphasis given to your own message and to the anticipated questions from the audience. In a scientific paper, you might concentrate on research results, whereas in a popular article or in a talk, the questions have more emphasis. Nevertheless, you should always adapt to the audience's prior knowledge in any type of communication.

SCIENTIFIC VERSUS POPULAR SCIENCE WRITING

Scientists usually communicate the same topic in various ways and to different audiences. New research results are often first communicated to other scientists at a conference, both in written form (sometimes just an abstract) and as an oral or poster presentation. The written conference paper normally follows the rules of scientific writing, but it is generally not reviewed by peer scientists before delivery.

The core method for communication of new research results is a paper published in a scientific journal. To meet the specific demands on clarity in communication of new knowledge, the scientific paper is organised in a standard way. The format normally used is called IMRAD, referring to the main sections of the paper: Introduction, Materials and Methods, Results And Discussion. The common practice for journals is to have the scientific paper reviewed by peer scientists, anonymous to the author, before it is accepted for publication. This is done to ensure that the results and interpretations published are of highest possible quality.

New findings in research also need to be communicated to public audiences, which means that, in addition to scientific communication, publishing and presenting popular science are necessary. Some characteristics of scientific and popular science writing are given in Figure 1:2.

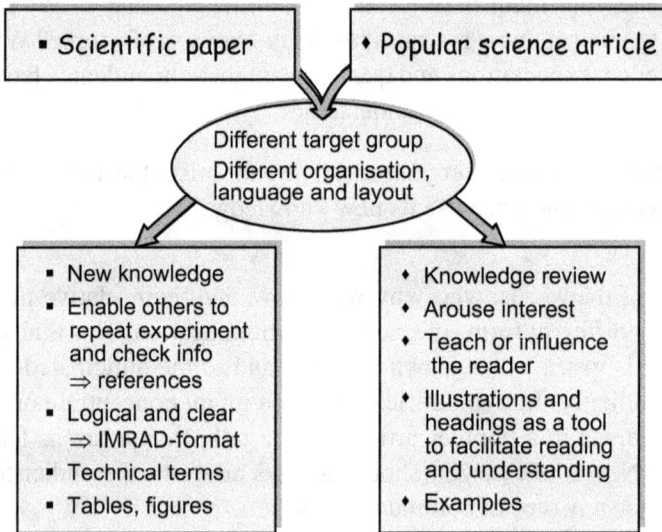

Figure 1:2. Some characteristics of scientific and popular science writing.

Scientific and popular science writing differ in several aspects, but there are also many similarities. The communication ABC, i.e. accuracy, brevity and clarity, should be fulfilled in each type of writing, of course, but the technical language cannot be the same for specialists as for non-specialists. The specific demands on scientific papers originate from the fact that original documentation of new research results requires precision. This precision includes distinguishing new results presented in the paper from previous results. Scientific writing should be logical and clear, so that the reader easily understands what is written with minimal misinterpretation, and so that others can repeat or verify what is done or said. In popular science writing, you first need to think about what in your research area might interest the reader, and second about how to explain things so they are understood, even without previous knowledge in this area.

Whether writing a scientific paper or a popular science article, it might be fruitful to have some characteristics of the other type in mind. Awakening interest and helping the reader is important, even when writing a scientific paper. Furthermore, reliable information ought to be a central feature, not only in scientific writing, but also in popular science writing.

THE SECTIONS OF A SCIENTIFIC PAPER REFLECT THE RESEARCH PROCESS

To understand the organisation of a scientific paper better, it is worthwhile looking at the major steps in the research process. These steps, as well as the corresponding sections of a scientific paper, are illustrated in Figure 1:3.

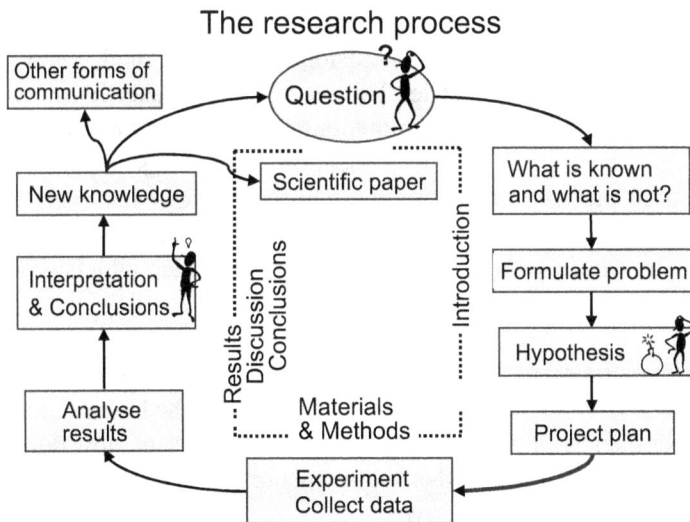

Figure 1:3. Major steps in the research process (outer circle) and corresponding sections of a scientific paper (inner circle). [1]

Research is exciting! We deal with something unknown, a search for an answer initiated by a question. We want to know why things happen, why

---

[1] After an idea of Backman, J. 1985. Att skriva och läsa vetenskapliga rapporter [Writing and reading scientific reports], p.17. Studentlitteratur, Lund.

there are differences, what relationships there are, how things can be explained. It is like looking for a missing piece to help solve a puzzle.

An important initial step in the research process is to find out what is already known about the topic by reading the scientific literature and by using other sources of information. If you do that, then it is possible to identify what is known and what is not known, and to formulate a specific problem to study. The research task will be determined by your hypothesis, which must be testable and formulated in such a way that you can either reject or accept it, depending on your results.

To test the hypothesis, it is necessary to collect data, e.g. by making observations, doing an experiment or a simulation study, using field data, or interviewing people. This requires a plan or experimental design, so that you know which materials and methods to use, and how the study should be structured to avoid confounding different effects in the interpretation of results. The results of the study are analysed, usually by statistical methods, and then interpreted and discussed in view of the previous knowledge, the hypothesis and the formulated problem. Conclusions are drawn, and a new piece is added to the knowledge puzzle. The piece might make the picture more complete or help to revise it. The outcome could be new recommendations and implementation, as well as identification of questions for future research. The research process thus forms a circle, where for each cycle the steps need to be documented and communicated to contribute to knowledge. The basis for this documentation is the scientific paper, which is normally supplemented with other forms of communication.

The sections of a scientific paper (the inner circle in Figure 1:3) reflect the research process, both in organisation and content. In the Introduction section, you answer questions such as, why is the topic important, how does it relate to previous knowledge, i.e. what is known and what is not known, and what are your hypothesis and objective?

The Materials and Methods section gives details on how you did the study, i.e. the project plan or experimental design, the materials used, the methods for making observations and the analysis of the data. The Results section describes what answers were obtained, often presented in the form of tables and figures. Results are discussed and interpreted in relation to previous knowledge, the formulated problem and your hypothesis, either in a separate Discussion section or in a joint Results and Discussion section. Conclusions can be included in the Discussion section, or given their own heading. The scientific paper will also include an Abstract (an independent summary of your paper) and a list of References, so that others will know where to find the literature referred to in your text.

Producing the scientific paper is an integral part of the research process. It is obvious, therefore, that thinking, planning and getting started in writing your paper can and should be initiated early in the process, especially because it helps to clarify your thoughts. Be aware, however, that writing takes time, and you will often have a deadline for delivery. Document continuously what you do, make notes of your ideas, organise the results you obtain in tables and figures, and set up a good reference system to keep track of the literature used. Don't wait for all results and analyses to be completed before you start writing your paper. You will find hints on how to deal with this topic in the chapter "Getting Started in Writing", but first we will discuss the sections of a scientific paper in more detail.

2

SECTIONS OF A SCIENTIFIC PAPER

During your professional career, you will be asked to prepare various types of written communication. Communicating science is as important to the scientific process as designing, conducting and analysing the experiment itself.

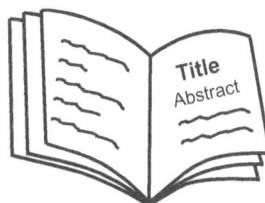

Journals publish different types of scientific papers, including original research papers; review papers; symposium papers; invited papers; technical research notes, which give results of complete but limited experiments; rapid communications, which deal with a "hot topic"; book reviews; and letters to the Editor, whose purpose is to discuss, critique, or expand on scientific points made in recently published papers.

MAJOR HEADINGS

Most scientific papers follow a standard format: Title, Abstract, Introduction, Materials and Methods, Results, Discussion, Conclusions, and References. Acknowledgements and appendices are optional, as are tables and figures. Sometimes the two sections, Results and Discussion, might be combined into one. Sometimes one section might be replaced by another (e.g. Theory might replace Materials and Methods in a paper that has a theoretical development) or even omitted (e.g. Materials and Methods or Results might be omitted in a review of the literature).

Major headings for review papers, conference papers, or short communications sometimes deviate from standard format. You might instead use section headings that are appropriate to the subject being reviewed. For all types of papers, follow the guidelines (often called Instructions for Authors) set by the journal or publisher to which you will submit the paper; journals usually publish these guidelines annually. The contents of standard major headings are discussed in the following sections.

Title

The title tells you what the paper is about, with the main purpose of encouraging people to read the paper. The title will be read more than any other section of the paper, so it should be informative, specific and concise. Make the title informative by describing the subject of the research, not results of the research. Make the title specific by differentiating your research from others on the subject. Make the title concise by limiting it to probably not more than seven to ten words. Journals give instructions to authors as to the length of the title. Put the most important words first. Write a working title first, but keep in mind that it might change as the focus of your paper changes while you write. Make sure that the final title accurately reflects the content of the paper.

The title should attract people to your paper

Only use technical terms if they are familiar to most readers and do not use abbreviations in the title. Eliminate "waste words", words that say nothing, e.g. "Observations of…", "Studies of…", "Investigations of…", "A note on…" or "Examination of…". Avoid titles with a series, e.g. I, II, III, …. Besides appearing self-aggrandising, which you should avoid, you might publish the first paper (I) but never publish another in the series (some journals insist that all papers in a series be submitted together). For review papers, consider using a two-part or "hanging" title, e.g. Xxxxx: A review.

If you are writing the title of an abstract for a presentation at a scientific meeting, remember that the title is often the primary basis for placing a presentation in a certain session of the scientific programme. Vague or uninformative titles increase the risk that the presentation might be allocated to an inappropriate session or might even be rejected. Beware of long titles that can look awkward on a title slide or poster.

The *running head*, which has a maximum number of characters plus spaces, is an abbreviated title and appears as a header on pages of the journal. Sometimes the journal specifies what the running head must be, depending on your paper. If, for example, you submit a short communication, the journal might specify the running head (e.g. SHORT COMMUNICATION:), followed by a short title. Make sure that your running head conveys the essential message from the title, particularly in a review paper for a book; it should attract people as they browse through the pages.

Abstract

The abstract is especially important because a large number of people will read it (e.g. through abstracting journals), and often it will determine whether they read the entire paper. Readers should be able to understand the abstract without having to be familiar with the details of the research.

Do you want your abstract to be indicative or informative? An indicative abstract indicates objectives of the research and suggests results in general terms. It makes the reader want to read your paper because it *might* be interesting. An

The Abstract can be read on its own

informative abstract, however, states the objectives and supports your conclusions with data. It makes the reader want to read your paper because it *is* interesting.

Whether you are writing an abstract for a paper or for a presentation at a scientific meeting, the abstract should describe the problem and summarise the major points of the research in a brief and understandable form. Begin with a clear statement of the objective. If space allows, however, start with the motivation and justification for research, then state the objectives. Continue with the approach and the main results, and end with one or two sentences that emphasise important conclusions and implications. Be specific and concise because the length of the abstract is often restricted to a maximum number of words or keystrokes (number of characters plus spaces), depending on the journal. The abstract should stand alone, i.e. can be read without having the full paper; so do not cite references to literature, tables or figures. Avoid using an abbreviation, but if you must, be sure that it appears in the journal's list of standard abbreviations, which do not require definition (e.g. DNA for deoxyribonucleic acid).

The abstract is usually written in the past tense for the objective, the materials and methods and the results, and in the present tense for the motivation and justification, the interpretation of the results, and the conclusions.

At the end of the abstract, list key words that best describe your research. Key words are used by indexing services and to form the subject index of the journal. Key words, therefore, should include when applicable the species, treatments and the major response criteria (see the chapter "Literature Searching and Referencing").

Introduction

The introduction persuades the reader that the topic is important and that the objective of the research is justified. It motivates and justifies the research, by providing the necessary background and by explaining the rationale for the study, and it clearly states the objectives and the approach. The introduction should orient the reader by summarising *briefly* the relevant literature.

Clearly state the objectives

Tell the reader what has been done and what has *not* been done; where is the gap of knowledge in the literature? Be specific and direct, but give the reader all necessary information leading logically to, and focusing quickly on, the objective and the approach. The introduction may also specify the question to be answered or the hypotheses to be tested.

Some journals want the introduction to include a brief preview of the materials and methods, e.g. the approach, and a brief preview of the results, e.g. the "take-home" message. Some journals want an extensive discussion of relevant literature to be in the discussion of results, not in the introduction. If you are writing a review paper, then the literature is not reviewed in the introduction. Review papers normally have a short introduction, a long discussion of literature (split into topics with headings), followed by conclusions; there may be a general discussion before the conclusions (see the chapter "Other Types of Scientific Writing").

Use the present tense to write the motivation and justification, as in "Labour is the largest expense in the production of ...". The review of literature, however, is written in the past tense, as in "Studies showed that ...", or in the present perfect tense, if it is common knowledge, as in "Studies have shown that ...". The objective is written in the past tense, as in "The objective was ...".

Materials and Methods

The materials and methods provide a clear and complete description for all experimental, analytical and statistical procedures. Organise the section logically, perhaps chronologically, and use specific, informative language. Include necessary information, and omit unnecessary information. It is not necessary, for example, to describe in detail a procedure already published; just cite the original reference but explain modifications. This section should include enough information so that another researcher can repeat the procedures and expect to get the same results. Use the past tense to write the materials and methods, as in "The treatment period was …".

The design of planned experiments, or the sampling protocol of surveys, should be conveyed clearly and concisely. In particular, the randomization, replication and blocking structure should be described to allow the reader to reconstruct the experimental layout. Describe clearly and fully your experimental subjects, numbers, treatments, environmental conditions, measurements and statistical models. Be sure to specify your experimental unit. State any assumptions you made, and state quantities in standard units. Microorganisms should be named by genus and species; specific strain designations and numbers should be used when appropriate. If you mention an enzyme, then you should include the EC number. For drugs, chemical names are more universally descriptive than trade names. For surveys, describe the sampling procedures and observational methods.

Consult instructions for authors if you refer to sources of products, equipment and chemicals used in an experiment. Model and catalogue numbers may be needed. Consult the journal also for specific instructions if you have sensory data or refer to computer software. It is not sufficient merely to indicate the software used to perform numerical computations, although such information may be useful, but also specify the mathematical model(s) and statistical methods used to analyse and summarise the data (see the chapter "Writing Statistics").

Results and Discussion

Results and Discussion sections may be separate or they may be combined. If separate, then the Results section should contain only results and a summary of *your* research, and not a comparison with the literature. If combined, then the Results and Discussion section should contain a comparison with the literature. Before starting to write about your results, it is best to prepare them in the form of tables or figures (see the chapter "Tables and Figures"). From among the tables and figures that you would like to report, select those that are the most important or the most representative – the ones that best tell your story.

The results section should explain or elaborate on your findings by logically summarising and illustrating relevant data, using the tables and figures you selected. Sufficient data should be presented to allow the reader to interpret the results. In the text, describe only the most important results that appear in a table or figure. Do not merely repeat numbers from the tables in the text, e.g. "The mean of X was 23.5 g", but rather integrate quantitative data into the text, e.g. "The mean of X (23.5 g) was greater...". If the text becomes too repetitive with similar results, just say that the results were similar. Means and standard errors are an efficient way to summarise in the text a large number of data in a table. Be as specific and informative as possible in describing results. Instead of writing, "X changed over time", for example, write, "X increased (or decreased or doubled or halved, as the case may be) over time" .

The Discussion section should interpret the results clearly, concisely and logically. Start by reminding the reader of the objectives. For each objective, in order, describe how your results relate to meeting that objective. Cite evidence from the literature that supports or contradicts your results; explain the contradictions. Identify the significant results, and recognise the importance of "negative" or non-significant results. Describe the limitations of your research, e.g. limitations with regard to the design of the experiment, the number of observations or the statistical analysis. Results, or references to tables or figures already described in the Results section, should not be repeated in the Discussion.

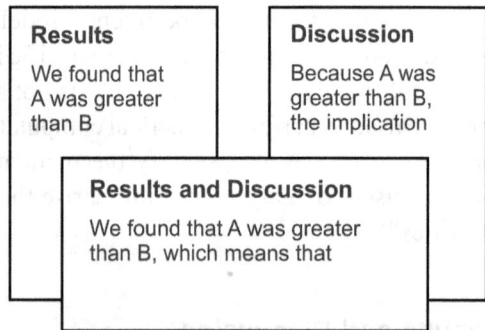

Results	Discussion
We found that A was greater than B	Because A was greater than B, the implication

Results and Discussion
We found that A was greater than B, which means that

Combining Results and Discussion can avoid repetition

If the journal permits, combining Results and Discussion into a joint section can be a good way to avoid repetition. Make sure, though, to indicate clearly whether you are reporting your own results or results from the literature. The main systems for citing references in text are presented in the chapter "Literature Searching and Referencing".

Use the past and present tenses to write the Results and Discussion sections: use the past tense to refer to your results, but use the present tense to interpret your results and to refer to results of others, as in "On average, A was greater than B, which means that …. This is consistent with NN (1999) who found that …" .

Conclusions or Implications

Conclusions summarise the main results of the research and describe what they mean in general. This is neither the place to mention new results for the first time nor the place to refer to the literature. Avoid abbreviations, acronyms or citations. Although some guarded speculation is allowed, you should avoid over-extrapolating the results. Provide readers with an interpretation of the impact of the research results, where appropriate. Consider suggesting future research to follow up where your research ended. Use the present tense to write the Conclusions, as in "Our results suggest that the direction for future research might be …".

Acknowledgements

Acknowledgements are often of two types: general and specific. General acknowledgements include those of an institution, of a laboratory, or of a source of funds, whereas specific acknowledgements include those of colleagues and technicians or of an anonymous (or even named) reviewer. If the research is part of your thesis or dissertation, then you might mention it here. If the paper contains a dedication, then this is where it might go. Some journals place acknowledgements as footnotes in the paper, so consult your journal's guidelines.

References

Citations of published literature in the text as well as in the reference list should follow the instructions set by the journal (or institution) where it is to be published. Instructions vary considerably, as discussed in the chapter "Literature Searching and Referencing". Be certain that all references listed are cited in the text of the paper, and that all references cited in the text are listed.

Appendix

The Appendix provides readers with supplementary material that might not be essential to the understanding of the paper but might be helpful. Such material could include numerical examples (which may instead be included in the text, depending on the nature of the paper), questionnaires, extensive details of analytical procedures, novel computer programs, or derivations of complex mathematical formulas or proofs. Alternatively, supplementary material may be provided on request from the author or on a web site; be sure the address of the author or the web site is available. Consult the journal for the correct placement of the Appendix section, which might vary by journal.

The main sections of a scientific paper can be summarised as follows:

Main sections of a scientific paper[1]	
Section	*Intends to tell the reader*
Title	What the paper is about
Abstract	Short summary that can "stand alone"
Introduction	The problem, what is known, what is not known, and the objective
Materials and Methods	What you did
Results	What you found
Discussion	How you interpret the results
Conclusions	Possible implications and the impact
Acknowledgements	Who contributed to the work and how
References	How to find the papers referred to
Appendix	Supplementary material

---

[1] For sections used in other written forms of science communication, see the chapter "Other Types of Scientific Writing".

3

TABLES AND FIGURES

Tables and figures, such as charts, diagrams, graphs, photos or other illustrations, help make numbers meaningful and convey information to the reader. They can be used to show relationships, to emphasise material and to present material more compactly and with less repetition. The number of tables and figures that you use depends on the amount of information you have and on your purpose in presenting that information – i.e. on the story you are telling.

The same data should not be presented in both tabular and graphical form; choose the form that best serves your purpose. Use tables when you want the reader to focus on specific numbers, e.g. actual data or estimates of parameters, and use figures when you want the reader to focus on relationships among those numbers. Consult the instructions of the journal on how to construct a table or a figure (e.g. the use of leading zeros or the symbols used for footnotes) and look at a recent issue of the journal for examples of format. Tables and figures are normally numbered in separate sequences. All tables and figures should be referred to in the text of your paper. Avoid sentences that only describe what is in a table or figure, e.g. "Figure 2 shows the change in colour during the experiment"; instead, emphasize the main result and put the reference in parentheses, as in "The solution got darker as time progressed (Figure 2)".

TABLES

Use tables to summarise numerical values so the numbers can be interpreted logically in the text of the paper (e.g. in the Results section). A table should not refer to the text (e.g. an equation in the paper) or to a figure or another table. A table should be self-contained, so as to stand alone. Every table must have a title that explains what the table shows. The title of the table should be descriptive enough not only to stand alone, but also to be used in the text to refer to that table. In the text, refer to the table using the same words as in the title of the table; then the reader is certain to be looking at the correct table. Use footnotes to make the table clearer and less "busy". Make the body of the table concise by avoiding repetitive information and by excluding data that can be computed from available information. Abbreviations can be used, but make them as "self-explanatory" as possible: if required, explain their meaning in a footnote. Use a logical format by arranging comparisons in columns (vertically) rather than in rows (horizontally) to facilitate mental subtraction or division (align columns by decimal point) and by emphasising similarities (group similar items) and differences (separate dissimilar items).

When possible, a table should be organised to fit the page, so that the page can be read without rotating it; orient the table as "portrait" rather than as "landscape". Avoid the use of vertical lines between columns and use few horizontal lines, except as needed (compare Tables 3:1, 3:2a and 3:2b).

You should make the table legible. Round data to simplify for the reader. Never use more significant digits than your method and data justify; for example, a mean of 503.753 kg implies that you can weigh to the nearest gram, which is not appropriate if your scales are only accurate to within 1 kg or 10 kg; in this example, the mean presented in the table should be either 504 kg or 500 kg. Use common units and double-space every five lines for easy reading. Avoid using a zero or a dash to indicate absence of results (unless your journal tells you to); zero is a valid observation, and a dash could be confused with a negative sign. Use "ND" to indicate "no data", "not determined", "not detected", etc. and explain ND in a footnote. Leading zeros (e.g. 0.113) make numbers easier to read and easier to justify in a column, but some journals omit them (e.g. .113).

If you write a report that is to be reproduced directly from your manuscript, then tables are usually inserted in the text. If your table is in landscape orientation, then rotate the page before putting it into the manuscript so that the bottom of the table is at the right-hand margin. A table should appear near the comment that refers to it. If a table will span a page break, then place it on the following page. When you submit a manuscript to be published in a journal, however, tables are usually put at the end of the manuscript, after the references. This is so the typesetter can format tables separately to suit the layout of the journal. You could indicate where you want them in the published paper by inserting the phrase "Table X near here".

Table 3:1. An example of a poor table layout

	Breed A		
Variable	Treatment 1	Treatment 2	Treatment 3
1 (units) 2 (units) 3 (units)			
	Breed B		
	Treatment 1	Treatment 2	Treatment 3
1 (units) 2 (units) 3 (units)			

Table 3:2a. An example of a better table layout

Treatment	Variable 1 (units)	Variable 2 (units)	Variable 3 (units)
Breed A			
1			
2			
3			
Breed B			
1			
2			
3			

Table 3:2b. Another example of a better layout (if more variables than treatments)

	Breed A			Breed B		
Variable	Treatment 1	Treatment 2	Treatment 3	Treatment 1	Treatment 2	Treatment 3
1 (units)						
2 (units)						
3 (units)						
4 (units)						
5 (units)						
6 (units)						
7 (units)						
8 (units)						

FIGURES

Use figures to focus on relationships among numbers. Use only as many figures as necessary to explain the results. Be aware that figures are often reduced in size for publication, which makes it difficult to distinguish between lines. Therefore, use an appropriate scale of lettering, symbol size and boldness of line; scale, size and boldness should be consistent across figures. Do not repeat material in a figure that already appears in a table.

Figures should be understood independently (stand alone), without reference to the text, to tables or to other figures. Figures with multiple parts, however, can be designated as 1a, 1b, etc. Abbreviations should conform to the style of the journal and be consistent with the text. Format and style should be consistent with other figures. Each figure must be accompanied by a caption or legend that explains what the figure illustrates.

Examples of figures include pie charts, bar charts, line graphs and scatter diagrams (Figure 3:1). Pie charts help to compare a part, or segment, with the whole. Start at 12:00 with the largest or most important segment and go clockwise in some logical order. Limit the number of segments to about five or seven. Label the segments outside the circle.

When comparing one item with another, use a bar chart or a line graph. Order bars in a logical way. Place all labels inside or outside the bars, not both. Unless you have three-dimensional data, do not use three-dimensional figures. This is especially true for bar charts, where three-dimensional bars are difficult to interpret and often obscure the presentation of data.

Line graphs help to compare items for responses over a continuous scale, such as time or levels of treatment, or to show frequency or distribution. Place time or levels on the horizontal axis. Avoid using more than three or four lines on a graph, if possible. To distinguish one line from another, use different markers, such as triangles, circles, or squares, or use different lines, such as solid or dashed. Do not used dotted lines, because they can appear solid when reduced for printing. Avoid using both different markers and different lines in the same graph. To show correlation between two variables, use scatter diagrams. Together with the line fitted to the data, scatter diagrams help to compare actual data with predicted data (regression lines).

When writing a report to be reproduced directly from your manuscript, figures are inserted in the text. If possible place a figure near the comment that

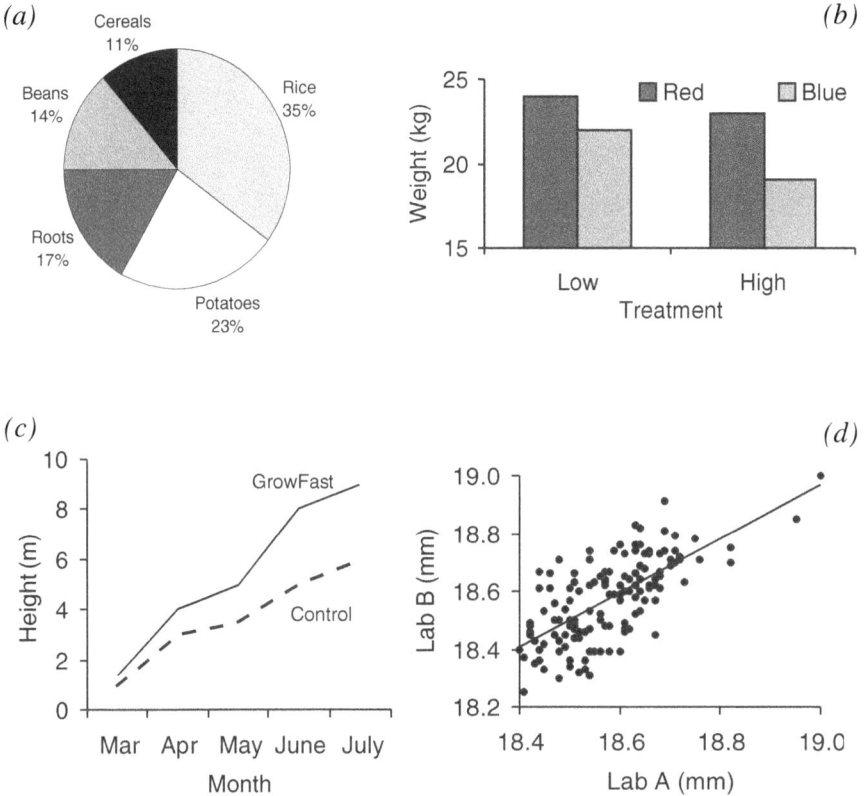

Figure 3:1. Examples of charts: (a) pie, (b) bar, (c) line, (d) scatter.

refers to it. If your figure is in landscape orientation, then rotate the page before putting it into the manuscript so that the bottom of the figure is at the right-hand margin. When writing a manuscript to be submitted to a journal, however, placement of figures may vary, so check the instructions; otherwise, figures are usually placed after tables. Be sure to identify each figure on the back with the figure number, the author's name, and the title of the paper; indicate the top of the figure, if necessary. Avoid the use of colour in figures, unless it is essential to understand the figure.

4

OTHER TYPES OF SCIENTIFIC WRITING

Scientific papers are the main method for communicating research results, but scientists and students have to produce many other types of writing. Examples described in this chapter are: literature review, conference paper and abstract, thesis or dissertation, popular science article and research proposal. The emphasis in each section is on how each type of writing differs from a standard scientific paper.

LITERATURE REVIEW

Good reviews contribute to scientific knowledge by bringing data together so that new, or more definite, conclusions can be drawn. A review differs from a standard scientific paper by reporting work from several sources, rather than from one experiment or research programme. Review papers are found in journals and conference proceedings; they are also a common form of writing in university training. Reviews of literature, furthermore, are found, in shortened form, in the Introduction section of a standard scientific paper and, in longer form, in the Literature Review section of a thesis or dissertation. The main requirement of a review is that it should be *critical*. This does not mean that you should criticise other authors, but rather that you should compare and contrast information published from different sources. Conclusions from a review of the literature might suggest new areas for research by identifying gaps in the knowledge. For advice on searching the literature, see the chapter "Literature Searching and Referencing".

The structure of a review paper is different

A review paper consists of an Introduction, various sections forming the body of the review, and Conclusions and References; there may also be a General Discussion before the Conclusions. The Introduction is similar to that of a standard paper: you state what the problem is and why you are going to review the literature, but you don't include the review in this section. The Conclusions and References follow the same principles as those in a standard paper. The major difference is in the body of the review.

The first task in writing a review is to split the body of your review clearly into themes or topics, each of which can have its own section and heading. Splitting your review into sections should help the reader to follow your reasoning. Keep each topic separate, but put them in a logical order. Starting with general topics and moving to specific topics usually works best, but you might relate specifics back to the general in the Discussion. In a review of hamburger consumption and human health, for example, you might have the following headings:

1   Introduction (what you are reviewing
    and why)

2   Influence of nutrition on human health

3   Nutrient requirements of humans

4   Trends in hamburger consumption

5   Nutrient content of hamburgers

6   General Discussion - hamburgers in a balanced diet

7   Conclusions

In this example, we move from the general situation (human health) to a specific situation (analysis of hamburgers) and tie everything together in the general discussion.

The content should reflect the contrasting results

In a scientific paper, you normally have a hypothesis, which you accept or reject on the basis of your experimental results. In a review, however, you have a theory or a message, which you support or contradict with published results. You will probably have formed the theory when reviewing the literature.

Even if you know the "answer" before you start to review the literature, you still need to convince the reader to accept your view with sound arguments, supported by good evidence. This applies to all sub-sections and statements in the review; do not simply make a statement that agrees with your idea and give only one reference. Try to have at least two references that support the idea. Even better, also try to give a reference that reports contradictory results and explain why it does not fit your theory (e.g. different experimental conditions). You should never omit a relevant reference merely because it conflicts with your ideas; equally, you should always be cautious when using references that agree with you but that might not be relevant (e.g. extrapolation beyond the data range).

Never extract parts of a paper that disagree with the original author's conclusions, unless you are looking at the data from a new angle. For example, if Smith (1995) wrote "We found that 99% of birds flew North, the rest flew South", you cannot write "Smith (1995) found that birds fly South". If some new studies found that brown birds fly North and white birds fly South, however, you might say "Colour appears to affect the direction of birds' flight (Brown et al., 2000), which might explain the observation of Smith (1995) that most birds flew North, whereas some flew South". Furthermore, do not report an author's results that were not significant as if they were.

Always try to give some experimental details from the paper that show how strongly the evidence supports your theory. For example, "In mice, … (Jones, 1999). Similar results have been found in rats, but only for females (Robinson, 2003)".

Quantitative data are useful for supporting theories and formulating new ideas. Rather than writing "Ingredient X increased fuel efficiency", write something like "Fuel efficiency increased from 10.5 km/l in control tests to 12.3 km/l when ingredient X was used ($P<0.05$)". Instead of describing every single study in the text of your review, it might be clearer to combine data from several studies in a table or figure. An example is given in Table 4:1. Be wary, however, of trying to do a statistical analysis on results from different studies; consult a statistician first.

Do not produce a long list of references to support a minor point; choose some key references that strongly support the point. Avoid referring to secondary sources, such as textbooks; they usually do not describe experimental results, they just state the author's opinions. When you get ideas or support from a paper that is itself a review, always check the original source of information; the reviewer might have misinterpreted the data.

Table 4:1. Example of table layout in a literature review (looking at the height of young people)

Sex	Number of observations	Age range (years)	Mean height (m)	Source
Male	250	10 - 17	1.84	Jones et al. (1990)
Female	250		1.76	
Female	100	11 - 13	1.68	Smith (1993)
Both	312	10 - 16	1.78	USDA (1995)
Male	96	12 - 18	1.85	Malik and Olsen (2002)

Writing your review

Like any other scientific writing, writing a review requires planning and careful thought. If you follow the advice given in the chapter "Literature Searching and Referencing", you will have lots of index cards (or their electronic equivalent) with key words on them. Sort your cards into the order of your review sub-headings. Within each topic, decide how you want to cover the material and sort the cards into the same order. Check whether you have any gaps (ideas without a reference) and fill them by doing a further literature search. Then gather the papers together in the correct order.

Write the review one section at a time. You may find that ideas can be moved from one section to another; choose which is best, but do not duplicate the same information in two sections. Check each section and make sure that you have evidence to support your arguments (both agreeing and disagreeing, if appropriate); check also that you have quoted each author accurately. When all sections are finished, read through the review and make sure that the flow between sections is coherent; you might need to add a sentence that links sections. For further hints, read the chapter "Getting Started in Writing".

Finally, write the General Discussion (if needed), the Introduction and the Conclusions. Make sure that the Conclusions refer to the Introduction, where you define the problem and objectives of your review. In the Introduction, for example, you might write "X is a potential problem"; in the Conclusions, you might write "X is a problem only in certain areas of the country ...".

CONFERENCE PAPER AND ABSTRACT

A conference paper may be in the form of a full scientific paper or a review, but you might be asked to write a separate summary or abstract. The difference between a summary and an abstract is that a summary is longer, so it can contain more detail, including tables and figures. Abstracts might be submitted in advance, so that the conference organisers can decide whether to accept your presentation and which session to put it in. Abstracts are often published as conference proceedings. The purpose of a conference paper or summary is to support your oral or poster presentation, so that you can concentrate on getting the main message across to the audience.

You will nearly always be given rules about the length and layout of your conference paper, summary or abstract. These rules should be followed carefully, particularly for an abstract, or your paper might be rejected. An abstract for a conference follows the same rules as an abstract for a scientific paper – concise, stand-alone, no references. Avoid using phrases such as "Results will be discussed", which give the impression that you have not yet done the research, and which are useless for people not attending the conference.

A conference paper or summary is usually written with the same section headings as a scientific paper (see the chapter "Sections of a Scientific Paper"), but you need not include as much detail in the Introduction and Discussion sections. You do not need a comprehensive review of literature, unless you are presenting only a review of literature. You should give just a few key references that directly relate to your work. The Materials and Methods section should be similar to that used in a scientific paper, possibly omitting sources of equipment, etc. The most important section of a conference paper or summary is the Results, which should contain full details of the results that you are going to present.

THESIS OR DISSERTATION

A thesis or dissertation can be produced for a PhD, MSc or BSc degree. The MSc or BSc thesis is usually a monograph, i.e. a complete "book" about your research programme. The PhD thesis, however, is often a monograph, but may, in some universities, consist of a compilation of published, submitted or draft papers, with a general introduction and a general

A thesis can be a monograph or compilation of papers

discussion to link the papers together. The main differences between the two approaches are style and layout, but some monograph theses look similar to a thesis based on a collection of papers. A thesis is often the most difficult piece of writing you will do in your scientific career. Try to spread the load by reviewing the literature and writing up experiments as you proceed through your research programme. See also the chapter "Getting Started in Writing".

The Introduction to a thesis or dissertation should set the scene and outline the approach adopted in the research programme. In some cases, an extended introduction is required, which includes a review of literature; in other cases, the Literature Review forms a separate section. Check the university regulations and follow the guidelines given in other sections of this book. The section "Literature Review" in this chapter should help. In an ideal thesis, you should conclude from the literature that there is a gap in knowledge; the reader then "turns the page" to see that your research has filled the gap.

The main body of a monograph thesis is normally split into chapters that correspond to individual experiments (like a collection of papers) or to different aspects of the research programme (more like a long scientific paper). If each chapter describes a separate experiment, then the sections on Materials and Methods, and on Results and Discussion might be included in each chapter. Alternatively, general chapters might be written for methods and discussion that are common to several experiments. The student should choose which layout to use, in consultation with the supervisor, but often the layout is determined by the nature of the research programme. There is no one correct way of dividing a thesis into sections.

The General Discussion for a thesis or collection of papers requires a slightly different approach from that for a scientific paper or review of literature.

You not only need to compare your results with previously published information and discuss the implications, but you also need to discuss the relationships among your individual experiments and to state how the overall programme fits your hypothesis. In other words, you must consider the whole picture as well as the individual pieces.

POPULAR SCIENCE ARTICLE

Research results need to be communicated not only to scientists, but also to public audiences, including those involved in the implementation of results. If this communication is not done effectively, then the new findings might not be spread and contribute to developments, and the public might feel anxiety and fear about what is going on in science research. Scientists, therefore, have great responsibilities in communicating their discoveries, but they often need training in how to do it effectively.

Research results should not remain embedded in the scientist's world

Writing a popular science article is different from writing a scientific paper, a review paper or a conference paper. Some characteristics of a popular science article were outlined in the chapter "Communicating Science" (Figure 1:2). We will now discuss in more detail what might be important to consider when writing a popular science article.

Adapt to your audience

Adapting to the readers' previous knowledge and experience is of utmost importance when writing a popular science article. Your writing will be different if the target group is directly applying your research findings (such as technicians or producers) than if the target group has no previous experience in the area. The readers must be able to relate what you say to their own world, or they will not understand your message; worse still, they will not even read the article. Analyse your aim and your audience; think of the questions Who? Why? What? How? when planning your popular science article (see the chapter "Communicating Science").

The subject matter affects your potential to attract readers, especially if you address a public audience. Your topic needs to be interesting per se, but it might arouse more interest if it is current and clarification is needed.

Simplify results – draw conclusions

Your popular science article should be focused on its topic. Therefore, give the general overview! Avoid presenting too many details, and make sure the information is put in context. Follow the communication ABC, i.e. be accurate, brief and clear, but give sufficient explanation for the reader to understand. Remember that the purpose is not only to communicate facts, but also to make them comprehensible and interesting, to discuss their possible application and impact, and to make the readers feel involved.

I didn't ask for all the details

One of the tasks in writing popular science is to simplify results; report only what is important for your message and omit the fine detail. This does not mean that you report only positive results; it is vital that you still give a balanced and honest report of your findings. In a scientific paper, you would give full details of statistical analyses, such as means and standard errors, whereas in a popular science article you might report only the means, or you might describe differences between treatments with only a few numbers. The process of simplifying results is similar to writing an abstract, except the words are rewritten for a popular audience.

In addition to simplifying the results, you should also omit most of the detail concerning materials and methods. Your experimental design, techniques and analyses will not interest the lay reader, and such information might detract from your message. Make sure to report your conclusions and implications, however, because these normally form the key message in popular science.

If you cannot decide how much information to include, whether your message is understood, or how to write something more simply, use a live audience. Try to explain your topic to a complete novice, such as your spouse, a relative, a friend, or a colleague from another discipline. You will have to use basic language to do this, so make your message simple; this will also give you a good foundation for building your article.

Organising the popular science article

A popular science article should have a structure that makes the reader want to read it immediately. Therefore, consider especially the:

- Title
- Preamble
- Headings
- Illustrations
- Layout

The title is of vital importance for capturing the readers' interest; so compose a title that is both exciting and informative. The title, usually in the present tense, must be short; it will often be a statement, but it might also be a question.

A popular science article often has some introductory text – a preamble – that is separated from the body text, to be easily visible. The preamble should make the reader curious and interested enough to continue reading; it should add some vital information, not just repeat what is in the title. The preamble is not like the abstract of a scientific paper, which summarises the main parts of the content; the preamble should emphasise the importance of the topic and hint about the contents of the article to raise curiosity.

The body text of a popular science article does not follow a standard structure; the text may be divided into sections, each having a heading. These headings are important tools for attracting readers. Headings make the text less compact, and they help readers to see quickly what the section (and the article) is about; they are like a road map! Create headings that are not only informative, but that are also eye-catching. Remember, however, that the headings should be short.

Illustrations can be effective in attracting readers, conveying a message and making complicated issues more understandable. Use photographs, drawings, clip-art, tables and figures (see the chapter "Tables and Figures"). Because illustrations capture readers' attention, the text in the captions can also be used to emphasise important messages. This text should explain what is shown in the illustration, but not repeat what is self-evident. The text might contain narrative, drawing some main conclusions from what is shown, for example, so that a reader who looks only at the title, headings and illustrations will still understand the essentials. Layout might be the main factor that encourages the reader to take a closer look at your article. The "house-style" of the magazine determines the final layout and length, but usually you can make suggestions. Remember that your text will more likely be read if it is not too long. Before writing your popular science article, look at the style of other articles in the same magazine.

Use language that is easily understood

Readers of a popular science article are usually not familiar with the language used in your scientific area. Use words that are easy to understand, therefore, and explain scientific or technical terms if you have to use them. You will find many ideas on how to make your language easier to read and understand in the chapter "Improving Your Writing".

Not everbody understands Greek

The tone of your writing can be more personal in a popular science article than in a scientific paper. What you say should stimulate the reader, and for that to happen your involvement with the topic is important. Use relevant examples and analogies to help the reader understand your topic. Furthermore, make logical transitions between sentences, paragraphs and sections, and check your text for "flow and thread", or coherence. See also the chapter "Getting Started in Writing".

RESEARCH PROPOSAL

Scientists spend a great deal of effort in writing research proposals, often with the purpose to convince a funding agency or other organisation to finance research projects. Sometimes also students write research proposals, e.g. as a first assignment in their graduate or postgraduate research project, although that proposal might not be an application for funding. Our discussion focuses on writing research grant proposals, but much of it is applicable also to student proposals.

A research proposal should be accurate, brief and clear. Competition for grants is usually high, so a research proposal has to be ranked "Very Good" to be considered for support. It should effectively convey why the proposed research is important and innovative, and also provide evidence that the applicant and possible collaborators are competent to do the job. In addition, the planned research should match the purpose and goals of the funding organisation.

Research grant proposals normally are read and graded by several reviewers. Some reviewers might be experts in the specific research area of the application, whereas others might not be. It is important when writing a research proposal, therefore, that it should be presented so that all reviewers can understand it easily. Reviewers usually read and evaluate a large number of applications in a short time, which further emphasises the need for proposals to be accurate, brief and clear.

Funding organisations normally have guidelines telling you what to include in a research proposal. Formal requirements, such as maximum number of pages with a specified font size and line spacing, specific application forms and number of copies to submit, are also specified. Make sure you follow the guidelines in every detail. If the formalities are not fulfilled, your research proposal might not even be evaluated!

Essentials of a research proposal

Writing a research proposal is similar to writing a scientific paper. You define the problem and the objectives, and you tell what is known and not known about the problem. You also present the research materials and methods to be used, as well as provide a time schedule and a detailed budget for the project. Instead of presenting results, you describe the expected outcomes. The qualifications of the applicant(s) are verified in a

Emphasise what is innovative in your proposed research!

"Curriculum Vitae". A research proposal, like a scientific paper, should be logical, and be structured with appropriate headings and an appealing layout.

When writing a research proposal, it is wise to check the criteria that will be used for evaluating the applications. Criteria commonly include:

- Relevance, i.e. significance of the proposed research activities in relation to the concerns and objectives of the funding organisation

- Scientific quality, e.g. originality, soundness of experimental design and research methods, realistic expected outcomes and multidisciplinary approaches

- Competence of applicants, research collaborators and organisations

- Dissemination and possible impact of research results

- Budget in relation to the research plan and funds available

Some essentials to include in a research proposal are:

Title. The title of your research proposal is what reviewers see first, so make the title informative, specific and concise, emphasising the main point(s) of your planned research. See also the advice given for writing titles in the chapter "Sections of a Scientific Paper".

Summary. Make an effort to write a good summary of your project; this is where reviewers get their first impression of the

Essentials of a research proposal

- Title & Summary
- Justification, background, objectives
- Research plan & Time schedule
- Expected outcomes
- Dissemination of results
- Budget
- Collaborating institutions
- Literature references
- Competence of applicant(s) - CV

planned research and of its importance. Summarise the key information of your proposal, e.g. the problem you want to address, the objectives, the significance and potential contribution of your proposed research, and, briefly, the methods you will use. The summary may also include a few words on your ability and your organisation's ability to carry out the research, as well as on the resources needed for the project. Guidelines for applicants normally state the length of the summary, which usually must fit into a defined space on the application form.

Justification, background and objectives. Define the problem and its background, and tell why your planned research is important and how it relates to the objectives of the funding organisation. Present a brief literature review (and a Reference List) to show what has been done already, and identify gaps in knowledge. Maybe you have some of your own preliminary results to present as well. Furthermore, clearly state your hypothesis and your research objectives, which might be split into overall and specific objectives. Remember to emphasise what is innovative and unique in your approach.

Research plan (including equipment) and time schedule. Relate the experiment or study to the objectives. Describe the research methods and materials to be used, including your experimental design and statistical methods for analyses of data. Note that ethical approval might be needed for some types of experiments. Justify your choice of research materials and methods. Describe the materials and methods so that the scientific quality of the proposed project can be evaluated, but avoid describing them in too much detail. State the facilities, equipment and other resources needed, what your organisation can provide and what will require funding within the research proposal. Present a time schedule for the activities to be performed and milestones to be achieved, e.g. as a time-delivery flow chart of achievements and outputs or "deliverables".

Expected outcomes. Specify the expected outcomes and possible applications of your research. The likely impact of the results might be discussed with regard to a specific sector, to society or to the development of the research area. Pay attention also to possible side-effects of the planned research, favourable and unfavourable, e.g. environmental impact and economic implications. The expected outcomes might be presented under a separate heading, or they might be described in the research plan in connection with each stage of the research.

Dissemination of results. Performing the research is not the only activity needed to meet research objectives; the scientific results must also be communicated to relevant audiences. In addition to publications in scientific journals and presentations at international and national conferences, the research results should also be communicated to the industry, to various agencies and to others outside the scientific community. Many funding organisations actually require a good plan for publication and dissemination of results to approve an application, but even without such a requirement it is wise to include plans in the research proposal for communicating the results.

Budget. Funding organisations often provide a specific budget format that you must follow, although it might be possible to add more details in the text of the proposal. The budget should be credible and realistic, and clearly reflect your research plan. Costs for salaries, travel, equipment, consumables and publications might be included, and there might be other costs to cover as well. Some items might need specific justification. Indicate whether your organisation, or maybe other organisations or donors, will cover part of the research costs. If that is the case, your chances for obtaining a research grant might be improved; cost-sharing/matching funds are sometimes a requirement.

Collaborating institutions. Performing the research in collaboration with other institutions might strengthen your proposal; collaboration might also indicate a multi-disciplinary approach, which is often considered beneficial. International collaboration sometimes is a requirement for funding; the budget might then cover the project costs for all collaborating institutions. Unless the institutions deliver a joint application, some written evidence of a planned collaboration should be provided with the application.

Curriculum Vitae (CV). The main purpose of the CV(s) is to provide reviewers with information to help them decide whether the applicant(s) have the competence needed to carry out the research described in the proposal. The CVs are often included as appendices to the research proposal. Each CV should be brief and clear; it should include only essentials relevant to the application. Organise the information into categories, such as personal facts, academic degrees, relevant positions, membership of professional organisations, awards or honours received, main research topics and relevant publications, as well as specific skills or experiences of importance for the research project. Don't overload the CV and harm your chances of success!

Before delivering a research proposal, read your draft carefully, and revise it for further improvement (see the section "Revise and edit . . ." in the chapter "Getting Started in Writing"). Read your proposal also with "eyes of a reviewer". Furthermore, ask some colleagues to read it, including one who is not in your specific research area. Finally, check that all requirements set by the funding organisation are fulfilled, including the appropriate signatures.

5

GETTING STARTED IN WRITING

Many of us feel tense and are resistant when the time comes to start writing. The task may seem overwhelming; our thoughts circle around and around. We try to deal with the document as if it were to be written in one single step. It is hard to see the structure and it is difficult to find the words. But don't despair; a research paper or report normally is not completed in a single step. Your writing can be divided into several stages, and thus can be made easier to handle:

- Analyse your aims and your audience
- Make tables and figures of interesting results, and decide what messages to communicate
- Make an outline
- Write a draft – start with the easiest parts
- Revise and edit

Whether you write a popular science article, a scientific paper, a report or a thesis, think of the questions Who?-Why?-What?-How? Who are your readers? Why is your topic important? What results do you want to communicate? How can you answer the most likely questions that might be raised by your readers? Having the readers in mind throughout your writing not only will make the text more interesting, but also will motivate you and make your writing more lively.

As you carry out your research, start writing early in the research process. Don't wait until all results are in or all analyses are performed. Put results together in tables and figures as you go along. Then you will get inspired, and you will know better what there is to write about. Starting to write early might even help you to identify missing parts in your research.

MAKING AN OUTLINE FACILITATES WRITING

Carefully reflect upon what to include in your document, and make a preliminary outline. This will help you to organise your thoughts, to structure the text so that it is logical and clear, and to remember important parts. The outline facilitates splitting the writing into steps and sub-steps, without losing the overview and the thread through it all.

Choose your own way to prepare an outline

An outline can be prepared in different ways. It might consist of a structured order of headings and subheadings, with key words under each, but it might also be written as an organisational chart or a mind map. The most important thing is that it should be a *working* outline, a tool to help you write. It is a plan for you, not for the reader. Your outline will not be complete from the start. It can, and normally should, be revised continuously during your writing. Once you have an initial outline, you can easily add key words or references to notes or literature.

Before starting to write your manuscript, discuss your proposed outline with other people, such as your co-authors, a colleague or, if you are a student, with your supervisor. This will help you to find the best approach and it will save time, because hopefully it will reduce the need for a large revision of the structure and content at a later stage.

Structure the text

The answer to the question, how to structure a text, is not simple. The structure differs depending on the type of text you are writing. Examples of structuring principles are:

- Chronological order (development over time)
- Order of interest/importance (the most important first)
- Cause and effect (or the opposite order)
- Comparison/contrast ("pros and cons")

When writing a scientific paper or report, you are normally compelled to include sections such as: Abstract, Introduction, Materials and Methods, Results, Discussion and Conclusions; within these sections, however, you can choose the structure. If you write a popular science article you are usually not tied to a specific format, but you need to structure the article so that the readers get interested immediately. Whatever kind of structure you use for your writing, make sure that it is logical and clear. Using different types of structures for different parts of a manuscript may sometimes be the best solution.

USE A COMPUTER FOR YOUR WRITING

Composing text directly at the computer is an efficient way to facilitate the writing process. Some of the main advantages are:

- You can easily revise text and layout

- You can easily merge text, tables, figures, etc. from various sources

- Knowing that it's easy to revise might have a positive "psychological" effect on your writing

Using word-processing doesn't change the need for logical thinking and clear writing, but it makes it easier to start writing, because your manuscript doesn't need to be "perfect" from the beginning. You can also use the computer to produce tables and figures you want to include in your manuscript. Sometimes you can do this within your word-processing programme, or you might need some other software for making graphs and drawings.

Valuable options in word-processing programmes

If you are not already familiar with the major options in your word-processing programme, it's worthwhile to learn them. It might also be useful to learn some short-cut commands, so that you don't need to mouse-click too often.

In word-processing, you can add or delete text easily. By using the *cut* and *paste* options you can move words, sentences, paragraphs, or entire sections, and you will see the result on the screen immediately. If you don't like the result of a change you can easily *undo* it. You can *search* and *replace* one word with another throughout the manuscript. Construction of tables can be facilitated by use of the *table* option. Numbers and text can be *sorted* in desired order. The *format* can be changed easily with regard to font type and size, bullets, indents, line spacing, margins, number of columns and page numbers. A *table of contents* can be produced automatically from your headings, and you can also produce an *index*. Advanced formulas and equations can be written with ease, although you may need additional software.

Spell-checking can be done for a large number of languages, either automatically or on your command. Be aware, however, that you still need to proof-read your manuscript carefully, especially for coherence. The spell-check doesn't function properly if a typing error has resulted in a word that exists, such as an error in the use of "to" and "too". *Grammar check* can also be helpful, but it sometimes disagrees with constructions that are common in scientific language, e.g. use of passive voice. Another useful tool is the *thesaurus*, which helps you find synonyms and antonyms. If you need to produce an abstract that stays within a specified number of words or characters, you might find the *word count* option helpful. You can also *compare documents* or *track changes* that are made in your manuscript by a co-author, for example.

The list of options in word-processing can be made much longer, and new options will be added all the time with new versions of software. The best way to learn the options is to test and practice, maybe read a book on the topic, or discuss with people who are experienced in using word-processing. In any case, remember to *save* your file frequently during the writing session, and to *back-up* your document on additional discs. It is also wise to make a printout.

START WITH THE SECTIONS YOU FIND EASIEST TO WRITE

Having produced some tables and figures, and also an outline, it is time to start writing. You know, or have an initial idea of, which sections and subsections you plan to include. You don't need to write your manuscript, therefore, straight through from beginning to end. It might be a better idea to start with the sections that you find easiest to write. This is true for most forms of scientific writing, although we will focus here on a scientific paper based on your research.

Writing a paper is like building a house

Writing a scientific paper can be compared to putting pieces together when building a house. Before you build the walls and rooms (materials, methods, results, tables, figures), you need a plan (outline) and a foundation (basic knowledge). The door and the windows (introduction and literature review) are produced separately and put in place later. The house is kept together by the roof (discussion), ending up with a chimney for the smoke (conclusions) to be spread widely. For a sound structure, the separate parts must be adjusted to fit together, just like logical transitions are needed throughout your paper. Once the house is in place, it is simple to make an overview picture of it (abstract).

Because you can build your scientific paper in pieces, it is easy to start planning or writing parts before the research is completed. Doing so means that you think of the paper early in the process. This may give ideas and thoughts that can continuously be put on paper, at least as key words or short notes. Writing takes time, so the sooner you start the process, the better!

You choose where to start

You might start to write a single section or you might write several sections in parallel; the choice is yours. When writing a single section, however, think of it as part of the full paper and of its role in the entirety. Two sections, the Materials and Methods and the Results, are often the first that are written. Some people find it easier to start with the Materials and Methods, because it is merely a description of what was done in the research. The first draft of that section can be written early in the research process. Reflecting on and writing about the materials and methods used might help to identify the need for supplementary or revised analyses. Other people, however, find it easier and more exciting to start with the Results. You can write this section using tables and figures that you have prepared from the analyses of your data. It might be a good idea to first explain your results to someone else; doing that can help you to get an overview of the results, to decide on what is important to include in the tables and figures, and to select those tables and figures to include in the paper. Once you have selected the appropriate tables and figures to help tell your story, elaborate on their contents in the Results (see the chapter "Sections of a Scientific Paper").

Having written a draft of the Materials and Methods and the Results, you probably will continue by writing the Introduction and the Discussion, either separately or in parallel, and then the Conclusions. The Abstract of the paper is written last.

WRITE A DRAFT – REVIEW AND REVISE

Before writing the first draft, it is a good idea to check the instructions for authors to find out the editorial rules and format to follow, including how to prepare tables and figures, and how to refer to literature in the text and in the reference list. Following the instructions from the beginning will save you time and frustration.

Get words down

In the first draft of your text, strive to get the words down. Don't feel pressured to make a "perfect" text; think of the content that you want to include, and write it! It is just a first draft, and you will revise and edit the text in future drafts. Imagine your readers asking you questions, such as: why is this important, what did you do, what did you find, how can you explain that, what use can we make of it and why are you telling me that? Remember to be accurate, brief, clear and logical in your writing. Emphasise those ideas that are most important, and develop them. What comes at the beginning of a sentence, a paragraph, a section or a chapter is usually considered to be the most important. Before you start to write, it might be worthwhile to have a look once again at the advice given in the chapters "Sections of a Scientific Paper", "Tables and Figures" and "Improving Your Writing".

Prepare for the next writing session

Before closing a writing session, write some key words on what you will write about in the coming part, and maybe even write the first sentences. Doing so will help you overcome the initial resistance you might feel to getting started next time. If you find it difficult to express your thoughts in writing, explain them to someone with the same background as your audience, so as to get feedback on what is understood. Then you'll find it easier to express yourself next time you sit down to write.

Revise and edit – let somebody read your draft

Revision of the text is usually done in several stages. The first revision might best be completed directly at the computer. Read through each paragraph and question what you wrote. Ask yourself "Am I saying what I mean?", "Would I understand this if it were new to me?" and "Is there a better way to say this?". Reflect on what you wrote. Think about ways to develop or to elaborate on your ideas, so as to tell your story better or to make the paragraph flow better.

Revise as needed to make the text clearer. Make sure, for example, that each pronoun (such as "it" or "this") refers to the appropriate noun. Then make a quick check of the text for fluency, for coherence and for typing errors. At that point, you might want to "step back" from the text for one or two days, so that you look at it with "fresh eyes" at the next revision.

In the second revision (probably best done from a printout) you might focus on the content and structure of your document. Does the title accurately reflect the content? Are all required headings included, and in the correct order? Will the abstract "stand alone", i.e. can it be understood without reading other parts of the paper? Is important content missing? Does the text agree with the tables and figures? Can the tables and figures be improved further to be clearly and easily understood? Is the source given where appropriate? Is the reference list in accord with the references cited in the text?

Having done this much revision, you would be wise to ask one or more people to read your draft (compulsory, of course, if you have a co-author(s) or a supervisor). This can be valuable and will help to get a view from other angles. It is recommended that you ask these people to read your draft once again, at a later revision of your paper. The more complete your text is, the easier it is to judge how well it can be understood and to see if something is missing.

At some later revision, we suggest that you read aloud your full text straight through from a paper copy. Make sure that the various parts fit together well - that there is coherence or a "thread" through it all. To improve coherence, put similar topics together. Check that the transitions between sentences, paragraphs and sections are logical. Delete repetition, unless it is necessary to emphasise an important message. Without losing the flow of your reading, indicate in the margin or directly in the text where revision is needed. Go back and do the revision.

When you feel that your text has "flow and thread", do yet another revision. Check for redundant words, long sentences, possible synonyms, etc. Make sure that your manuscript follows the editorial rules and format of your publisher. Spell-check after every revision; even better, let the function be on continuously. Do a final check that references in text and in the reference list agree; you might have added or deleted references during your revision.

If you write in a foreign language, a linguistic review by a native-speaker might be the last step in preparing your manuscript. Furthermore, if you are writing a monograph thesis or a report, remember to update the Table of Contents when all your contents are in place.

6

IMPROVING YOUR WRITING

Vigorous writing is concise. A sentence should contain no unnecessary words, a paragraph no unnecessary sentences, for the same reason that a drawing should have no unnecessary lines and a machine no unnecessary parts. This requires not that the writer make all sentences short, or avoid all detail and treat subjects only in outline, but that every word tell.

We hope that this chapter will make it easier and more enjoyable for you to write a clear and concise scientific paper. The examples given are mainly applicable when writing in English, but most of the principles illustrated are useful also when writing in other languages. One note of caution, however; many recommendations given here pertain to British English, and not necessarily to U.S. English, so be sure to follow the instructions to authors given in the journal to which you will be submitting your paper. Many of the examples in this chapter are taken from Locker (2003) and further help can be obtained from the books and articles listed in the chapter "Further Reading". You can find advice for writing and revising your text in the chapter "Getting Started in Writing". We also recommend that you have a good dictionary and grammar book close to hand.

HOW TO MAKE YOUR WRITING EASIER TO READ

Writing a scientific paper can be easy and even fun. One way to make it easy is to decide on a set of rules that will guide your writing style and then adhere to those rules. You might decide, for example, that *"if"* should be followed by *"then"*, as in the sentence: *"If* I follow my rules as I write, *then* many of the decisions about style are already made". Your rules may change over the years, as you improve your writing style; that is OK, and even desirable. If writing is easy for you, then your writing will be easy to read. We will give hints on how to make your writing easier to read.

Choose words carefully

Use words that are accurate – that mean what you want to say; words that are appropriate – that fit well with other words in the paper; and words that are familiar – that are easy to read and understand. Use specific, concrete words; they are easier to understand and remember.

When you have something simple to say, say it simply. Use the word that conveys your meaning most accurately, but when deciding between two such words, choose the shorter word.

For example:

Instead of ...	*Use ...*
approximately	about
commence	begin
finalise	finish
prioritise	rank
terminate	end
utilise	use

There are exceptions to this rule, however. Use a long word if it is the only word that expresses your meaning accurately, if it is more familiar than a short word, if its connotations are more appropriate, or if scientists in your discipline prefer it. Use technical words and expressions (jargon) only when the terms are essential and familiar to the reader. Otherwise, avoid the use of jargon because it is often difficult to understand. Instead, use a simpler "plain-language" equivalent, even when the equivalent expression is longer.

Use active verbs and avoid passive verbs

A verb is *active* when the subject does the action. A verb is *passive* when the subject is acted upon. To identify a passive verb in a sentence, look for a form of the verb *"to be"* followed with *"by"*. For example, "This method *was* recommended *by* them" is passive, whereas "They recommended this method" is active.

A sentence that uses an active verb is shorter and clearer, more interesting and less boring, more direct because it emphasises the subject, more forceful, more efficient because it takes less time to read and is easier to understand, and sounds less pompous and bureaucratic.

There are situations, however, when it might be desirable to use a passive verb: to emphasise the receiver of the action rather than the one doing the action, e.g. "Individuals were measured weekly"; to avoid assigning blame, e.g. "It is known that there are errors associated with field data"; or to omit an unknown or irrelevant person or thing doing the action, e.g. "Data were analysed …".

Passive verbs are desirable also to provide coherence within a paragraph, i.e. to provide transitions between sentences by repeating a word. A paragraph might be easier to read if you use the "new – old" construct: end one sentence with "new" information and begin the next sentence with "old" information, even if it uses a passive form. For example, "The epidemic ended with the discovery of a vaccine {new idea}. The vaccine {old idea} was developed by ...".

Use strong verbs – not nouns

Put the emphasis of the sentence in the verb. Strong verbs make sentences more forceful and easier to read. Instead of writing "We performed an analysis of the data", where the action word is the verb "performed", but where the important word is the noun "analysis", it is more forceful to write "We analysed the data", where the action word is the verb "analysed". Nouns ending in -*ment*, -*ion* and -*al* often hide the verb.

For example:

Instead of using the noun in ...	*Use the verb ...*
make an adjustment	adjust
make a judgment	judge

Instead of using the noun in ...	*Use the verb ...*
make an assumption	assume
reach a conclusion	conclude
take into consideration	consider
make a decision	decide
perform an investigation	investigate
make a selection	select
make a referral	refer

Tighten your writing

If an idea can be expressed in fewer words, then the writing is wordy. Wordy writing bores the reader and makes it harder to understand what you mean. Good writing is tight, and tight writing is good, because it allows you to convey more information. Tight writing is important, especially when you have to write an abstract with a limit on the number of words or keystrokes (characters plus spaces).

To tighten your writing, follow these strategies:

- Eliminate redundant words whose meaning is already clear:

  a period of two months, during the course of the experiment, during the year 2002, maximum (minimum) possible, past experience, plan in advance, refer back; the colour red, true facts, repeat again, already existing, different alternatives, previous literature, completely eliminate.

- Delete words that don't add to the understanding of the message:

  it is interesting to note that, it should be pointed out, it is significant that, the (especially with plurals, e.g. "Data were analysed . . ." means the same as "The data were analysed . . .", so delete "The").

- Delete words that can't be quantified when used to modify another word:

  quite, really, rather, very, many, few.

- Substitute a single word for a wordy phrase:

Instead of using ...	*Use ...*
at the present time	now
due to the fact that	because (NOT since)
it may be that	perhaps
in the event that	if
in the near future	soon
prior to the start of	before
on a regular basis	regularly
the first point is	first
more often than not	usually, normally
would seem to suggest	suggests
one of the problems	one problem
in spite of the fact	although, despite, nevertheless
on two separate occasions	twice
in close agreement with	agree with
take into consideration	consider
carry out experiments	experiment
it is obvious that	obviously
the majority of	most.

- Use the infinitive (the *to* form of a verb; e.g. to run), not the gerund (the *-ing* form of a verb; e.g. running), to make a sentence smoother and shorter.

- Combine sentences to eliminate unnecessary words and to focus attention on key points.

- Put the main idea of your sentence into the subject and verb to reduce the number of words. Instead of writing "The purpose of this approach is to allow ...", try writing "This approach allows ...". The main idea is not *purpose* but rather *approach*.

Think about what you actually *mean* to say, write it in different ways, and choose the tightest one.

Phrases beginning with *of, which* or *that* often can be shortened. Instead of writing, for example, "The estimates of the parameters were ...", it is shorter to write "Parameter estimates were ...". Sentences beginning with *It is* or *There are* can often be tightened. Instead of writing "There are three reasons for these results", it is tighter and stronger to write "The three reasons for these results are ...".

Keep sentences short and simple

Simple sentences have one main idea. Compound sentences have two main ideas that are closely related and joined by conjunctions, such as *and*, *but* or *or*. Complex sentences have one main idea and one subordinate idea related logically, e.g. "If ..., then ...".

Strive to put the most important idea early in the sentence. Instead of starting a sentence with a reference to some research, start with the main finding of the research and place the reference at the end of the sentence.

Always edit sentences for tightness, but use long sentences to link ideas, to avoid a series of short, choppy sentences or to reduce repetition. When using long sentences, keep the subject and verb close together by putting "qualifying" material at the beginning or the end of the sentence. Instead of writing "The mean for Treatment A was ...", write "For Treatment A, the mean was ...".

You can vary sentences by changing the order of the elements of the sentence, not by changing the elements. If, for example, you start to use "experiment" to describe a situation, don't replace it by "study" or "trial" in the subsequent text; continue to use "experiment", otherwise it is confusing to the reader.

Avoid using "respectively" in your writing. Instead of writing "Results for Treatments 1, 2 and 3 were X, Y and Z, respectively", it is easier for the reader if you place together elements that go together, so write "The result for Treatment 1 was X; for Treatment 2, Y and for Treatment 3, Z". In that way, the reader has to read the sentence only once.

Some journals suggest that you avoid writing *and/or*, because it appears as if you can't decide which is appropriate. Decide if the appropriate meaning is *and*, as in "X and Y", or *or*, as in "X or Y" or in "X or Y or both".

Use parallel construction

Use the same pattern for ideas that have the same logical function. Balance elements of the sentence: nouns with nouns, verbs with verbs, adverbs with adverbs, and prepositions with prepositions.

Parallel construction makes writing smoother, more forceful, and easier to understand. Use parallel construction for items in a list or for a series of ideas, for comparisons using *than* or *as,* or for contrasts using *however* or *whereas*. Parallel construction is especially useful when writing the results and the discussion. Whole sentences can parallel whole sentences, and whole paragraphs can parallel whole paragraphs. You can write, for example, "For Group 1, Variable X was lower for individuals on Diet A than on Diet B, whereas Variable Y was higher for individuals on Diet A than on Diet B. For Group 2, however, Variable X was higher for individuals on Diet A than on Diet B, whereas Variable Y was lower for individuals on Diet A than on Diet B". Once you decide on the pattern to present the result for one variable, use the same pattern to present results for other variables; simply copy, paste and revise the text. Having understood the result for one variable, the reader can see the pattern and can understand results easily for other variables.

Link ideas with transitions

Transition words and phrases (e.g. *and, for example, whereas, during*) signal connections between ideas. Proper transitions are key factors to make a text easy to read. Transitions tell the reader if the next sentence continues the previous idea or starts a new one. They tell the reader whether the idea that comes next is more or less important than the previous one. Transitions are used also to introduce the last or the most important item, to introduce an example, to compare or contrast ideas, to show cause and effect, to show time, or to summarise or end. The following are examples of transition words and phrases:

* To show continuation of an idea:
    and, in addition, also, likewise, first, second, third, similarly

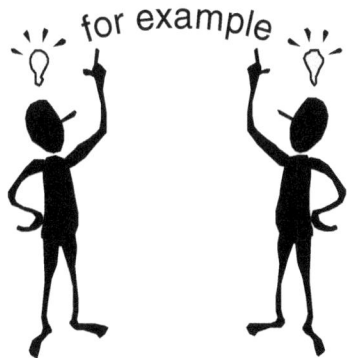

- To introduce an item:

    finally, moreover, furthermore

- To introduce an example:

    for example (e.g.), to illustrate, for instance, namely, indeed, specifically, on the one hand

- To contrast:

    in contrast, on the other hand, or, whereas

- To show that one idea is more important than another idea:

    but, nevertheless, however, on the contrary

- To show cause and effect:

    as a result, for this reason, because, therefore, consequently

- To show time:

    after, next, as, then, before, until, during, when, in the future, while, since

- To summarise or end:

    in conclusion, to summarise

Write coherently

Coherence refers to the logical sequence of sentences within a paragraph. Just as you should begin a sentence with the subject, you should begin a paragraph with the topic sentence, which states the main idea of that paragraph and provides a unifying structure to the paper. A good topic sentence forecasts for the reader the content of the paragraph and holds the paragraph together. If, for example, your topic sentence were "Labour is the major cost of production", then the reader would expect the paragraph to be about labour. If, however, your topic sentence were "The major cost of production is labour", then the reader would expect it to be about cost of production. A paragraph that lacks a topic sentence lacks unity. If a paragraph contains more than one main idea, consider linking ideas with a transition sentence. Otherwise, consider splitting the paragraph into more than one.

To improve coherence within a paragraph, discuss only one idea, or one topic, at a time. Use the "new-old" construct to move forward with your idea. Use the same organisational pattern for successive sentences by using parallel structure for the main subjects and main verbs. Tell the reader what to expect, e.g. "Two problems with this method are X and Y"; then go on to elaborate on the two problems.

Write logically

Write what you actually mean to say and write it logically. If you have difficulty putting an idea onto paper, e.g. because of "writer's block", say to yourself "What I actually mean to say is ...", and then write the words you mean to say. Once the words are on paper, you can revise them. Above all, make sure what you write is coherent and makes clear, logical and scientific sense. Using transition words or phrases will help make the logic flow and move the reader forward in the paper to its logical conclusion. It is important to organise a scientific paper according to the standard format for the journal. It is also important to write the paper in a style that uses clear and concise language, so as to allow the reader to understand your research paper thoroughly and efficiently.

Finally, when you have finished writing the paper, REVISE, REVISE and REVISE.

DO I OR DON'T I?

As you improve your writing skills, you will develop rules about style of writing. Some rules about style are always true, e.g. that you start a sentence with the main idea or that a subject and its verb must agree in number, whereas some rules are only partly true and should be applied carefully. For those "rules" that are only partly true, you might ask, ***Do I or don't I ...***

- ***...split an infinitive?*** The infinitive form of a verb is the "to" form, e.g. "to understand". Avoid splitting the infinitive with an adverb. Instead of writing "to better understand the results ...", write "to understand the results better ...". This rule of style is not "hard and fast", however, so split the infinitive if it is necessary to make the sentence easier to read.

- ***...use the first person* (I *or* we)?** Using *I* or *we* when you write about things that you did in the Materials and Methods might be more appropriate and smoother than using passive voice or using awkward phrases, such as *this writer* or *these authors*. Some journals, however, might object to this style of writing, whereas others might not. Some journals might even suggest new wording, e.g. "Our objective was ..." instead of "The objective was ...".

- ***... begin a sentence with* and (also) *or* but (however)?** If you begin a sentence with *and* or *also*, then it appears to the reader that the next idea is an afterthought. Instead, embed *also* at an appropriate place in the sentence; *in addition*, *moreover*, *therefore* or *furthermore* can also be embedded. If you begin a sentence with *but* or *however*, then it tells the reader that the next idea is in contrast to the previous idea. It is better to embed the word in the sentence, however, so as to begin the sentence with the most important idea, the subject.

- **...*end a sentence with a preposition*?** The end of the sentence, like the beginning, is important and should be emphasised. Prepositions, e.g. *at, for, in, under*, and *with*, are usually not worth emphasising and, therefore, should not be at the end of the sentence. One expects something to follow the preposition, but nothing does when the preposition is at the end. Choice of preposition is sometimes difficult so, when in doubt, try "*for*"; it almost always works. If, for example, you cannot decide whether to write "the variance *of* X" or "the variance *in* X", then try "the variance *for* X" and stay with it. It does not help the reader if you vary the construction.

- **...*impress my readers with big words*?** The purpose of writing a scientific paper is to inform your readers, not to impress them or to show off that you know big words. If you try to impress your readers with big words, then not only does it make your writing sound pompous, but it also makes you look foolish if you used the big words incorrectly. For example, "The postulation that generously proportioned verbiages make an affirmative impression on personages who inspect your inscriptions is a fallacy" means (almost) "Big words do not impress readers". [Note that some of the big words in this example are used incorrectly]. Big words used correctly, however, can convey information effectively. Don't hesitate to use a big word or expression, therefore, if it conveys the appropriate meaning clearly and concisely.

WRITING CORRECTLY

Everyone, including native speakers of English, makes errors while writing in English. For effective writing, it is best to avoid making errors. Common errors involve issues of spelling, grammar, punctuation and word choice. Errors of spelling can be corrected easily using spell-checking software, but there are dangers. If the error spells another word, then it will not be corrected; if you use the wrong computer dictionary (especially U.S. versus British),

Use the correct dictionary

however, then you might introduce errors instead of correcting them. Errors of grammar, punctuation and word choice, are more difficult to correct (grammar-checking software may give good suggestions, but often cannot cope with scientific writing). Examples of common errors, and some suggestions for avoiding them, are given below.

Grammar

Grammar is one aspect of writing that writers seem to find most troublesome. Common issues of grammar include:

Agreement

Be sure that subject and verb agree in number (singular or plural); a singular subject takes a singular verb, whereas a plural subject takes a plural verb. Use a singular verb when singular subjects are joined by *or*, *nor*, or *but*, e.g. "Neither Treatment 1 nor Treatment 2 was …". Use a plural verb when two or more singular subjects are joined by *and*, e.g. "Treatments 1 and 2 were …". When the sentence begins with *Here* or *There*, be sure the verb agrees with the subject that follows the verb, e.g. Here is one … or There are several …. Some words that end in *s* (e.g. *series*) are singular and require singular verbs (e.g. *is* or *was*). Some words that do not end in *s* (e.g. *data*), however, are plural and require plural verbs (e.g. *are* or *were*). When proof-reading your paper, find the subject and the verb, and make sure they agree in number.

Be sure also that the noun and pronoun agree in number. Words that take a singular pronoun include: *each, either, everybody, everyone, neither, nobody* and *none*, e.g. "Each of Treatments 1 and 2 was …". If a situation does not follow the rule or if the rule results in an awkward sentence, however, revise the sentence to avoid the problem. *None*, for example, can take a singular or a plural verb, depending on the context of the sentence. If it is treated as singular, then use a singular verb; if it is treated as plural, however, use a plural verb. The word should not cause you to stumble during the sentence when you read your paper aloud.

Each pronoun must refer to a specific word, without ambiguity. If there is any doubt which specific word the pronoun is referring to, then add the word in the sentence to make it clear and unambiguous. Make sure especially that *this, these, they* and *it* refer correctly to a specific antecedent (a word or phrase to which another word refers). If it is not clear which antecedent is example, "This suggests that ...", write, "This result suggests that ...". Use *who* or *whom* to refer to people, e.g. "the people who" not "the people that". Use *which* to refer to things, but use *that* to refer to animals, organisations or things.

Dangling modifier

A modifier is a word or phrase that describes the subject, verb or object. A *dangling* modifier, however, is a word or phrase that describes a word that is not in the sentence. Don't write, "While performing the surgery, the patient was anaesthetised", when you mean to write, "While the doctor was performing the surgery, the patient was anaesthetised". Note that the doctor, not the patient, is doing the surgery, but the doctor is not in the original sentence. When using a verb or adjective that ends in *-ing*, like *performing*, be sure it modifies the subject of the sentence, *doctor* not *patient*.

Your choice of words
is important!

Misplaced modifier

A *misplaced* modifier modifies an element of the sentence other than the one intended. To correct the misplaced modifier, therefore, move it closer to the word it modifies or add punctuation, if necessary, to clarify the meaning. One word that is often misplaced is *only*. Instead of writing, "We only measured two samples", write, "We measured only two samples". Other words that are often misplaced include *probably*, *properly* and *usually*.

Parallel structure

Items in a series or list that are the same in content and form must have the same grammatical structure; it is not easier for the reader if you vary the construction. The same construction, or parallel structure, makes it easier for the reader to recognise the similarity of content and form. If the sentence is short, then the article or preposition that applies to each item in the series may be used only before the first item. Instead of writing, "Tissue samples were weighed, then frozen and analyses were performed", write, "Tissue samples were weighed, frozen and analysed". If the sentence is long, however, the article or preposition should be repeated before each item; for example, "Tissue samples were weighed to the nearest ..., were frozen at ... and were analysed by ..."

the article or preposition should be repeated before each item; for example, "Tissue samples were weighed to the nearest …, were frozen at … and were analysed by …"

Correlated expressions, such as *both…and…*; *not only…but also…*; *not…but rather…*; *either…or…*; *neither…nor…*; *first…, second…, third…*, must be followed by the same structure, as in "Results suggest not only that …, but also that …".

Predicate errors

The predicate of a sentence, the part that refers to the subject, must fit with the subject. In sentences using *is* or other linking verbs, the complement must be a noun, an adjective or a dependent clause. Instead of writing "The reason for this result was *because* …", it is better to write "The reason for this result was *that* …". Be sure that the verb describes the action done by or done to the subject. Instead of writing "mothers were weaned at about six months", it is better to write "infants were weaned at about six months", because weaning describes the action associated with the infants not the mothers.

Punctuation

Punctuation helps the reader know what to expect. Punctuation takes the reader through the sentence and from sentence to sentence. Use correct punctuation, as in the following examples, to help avoid common errors in writing.

Comma splices and run-on-sentences. A comma splice occurs when two independent clauses are joined only by a comma, and not by a comma and a coordinating conjunction (e.g. *and, but, or, nor, for, so*). For example:

☹ The experiment began in June, the exact date was unknown.

Correct a comma splice by using a comma and adding a conjunction. A comma splice can also be corrected by using a semicolon, if the ideas are related, or by starting a new sentence, if the ideas are not related. For example:

☺ The experiment began in June, but the exact date was unknown.

☺ The experiment began in June; the exact date, however, was unknown.

In a run-on sentence, the independent clauses are joined neither by a comma nor by a conjunction. For example:

☹ The experiment began in June the exact date was unknown.

Correct a run-on sentence by adding a comma and a conjunction, by using a semicolon, if the ideas are related, or by separating it into multiple sentences.

☺ The experiment began in June, but the exact date was unknown.

☺ The experiment began in June. The exact date, however, was unknown.

Sentence fragments. A sentence fragment is not a complete sentence, but is punctuated as if it were. For example:

☹ By monitoring the rainfall, we measured two variables. The amount in mm. Also, the duration in hours.

☺ By monitoring the rainfall, we measured two variables: the amount in mm and the duration in hours.

☺☺ For each rainfall, we measured the amount in mm and the duration in hours.

Apostrophe (') Use an apostrophe in a contraction to indicate that a letter is omitted, e.g. *it's* for *it is*; to indicate possession, e.g. the person's DNA or the students' notes; or to make a plural, but only if it might be confused with another word, e.g. A's, so as not to confuse it with As. With dates, 1990s means the decade 1990–1999, whereas 1990's means belonging to the year 1990.

Colon (:) Use a colon to separate a main clause from a list or to join two independent clauses when the second clause explains or restates the first, as in "Two variables were measured: intakes of sugar and water".

Comma (,) Use a comma to separate the main clause from an introductory clause, or words that interrupt the main clause. Use a comma also after the first clause in a compound sentence, if the clauses are long or if they have different subjects. Use a comma to separate items in a series, but consult the journal instructions to learn if you should place a "serial" comma before the *and* or *or* for the final item. Do not use a comma simply because you stop to take a breath.

Dash (—) Use a dash to emphasise a break in thought. Don't confuse a dash with a hyphen; a dash is two unspaced hyphens.

Ellipsis (...) Use an ellipsis (three dots) to indicate that one or more words have been omitted, usually within quoted material.

Full stop **or** ***period*** (.) Use a full stop at the end of a sentence. Use a full stop after Latin abbreviations (etc.), but not after metric measurements (kg), chemical symbols (HCl), acronyms (BBC), or abbreviations where the last letter is the same as the last letter of the full word (Dr), especially in British English. Note that practices vary regarding full stops in abbreviations, so consult your journal instructions.

Hyphen (-) Use a hyphen to join two or more words used as a single adjective, e.g. two-day-old baby [NOT two-days-old baby] or two day-old babies.

Parentheses **or** ***brackets*** ( ) Use parentheses to set off words, phrases or sentences used to explain the main idea, e.g. "Maximum daily gain (10.5 kg) was achieved by ...".

Quotation marks ("" **and** ' ') Use quotation marks around words that you are discussing as words, e.g. "The answer 'Yes' was most frequent", or around words that you quote directly from someone else. Limit quotation marks to three sets, if possible, e.g. double marks outside, single marks in the middle and double marks inside, as in "...'... "...".'...". Journals have different rules about single (' ') or double ("") quotation marks, so check their instructions. In U.S. English, put a full stop (.) or a comma (,) inside quotation marks, but put a colon (:) or a semicolon (;) outside. In British English, all punctuation marks are placed inside quotation marks if they are part of a quote, but outside quotation marks if the quote is part of a longer sentence.

Semicolon (;) Use a semicolon to join two closely related independent clauses or to separate items in a series when the items themselves contain commas.

Slant line **or** ***slash*** (/) Use a slant line to mean *per* with units of measurements and *divided by* in equations. Use only one slant line in an expression or it becomes mathematically ambiguous; e.g. instead of "ppm/ha/yr", write "ppm/ha per year".

Square brackets ([ ]) Use square brackets to make additions or comments to quoted material not a part of the original quote, e.g. [sic], or to provide explanatory information. [sic is the Latin word for "so" and is used to indicate that you have quoted something exactly as it was written, no matter how absurd or incorrectly spelt].

Words that are often confused

Some English words in scientific writing are often confused. Master the following examples and you will be able to use them correctly in the future. If you are not sure of their correct usage, consult an English dictionary.

accept / except
> accept: receive
> except: leave out or exclude, but

access / assess / excess
> access: the right to use, see or enter
> assess: make a judgement, calculate
> excess: surplus

advice / advise
> advice: (noun) counsel
> advise: (verb) give advice

affect / effect
> affect: (verb) influence or modify
> effect: (noun) result

affluent / effluent
> affluent: (adjective) rich
> effluent: (noun) something that flows out

aggravate / irritate
> aggravate: make a bad situation worse, add to
> irritate: annoy, vex

a lot / allot
> a lot: many
> allot: divide, give

allude / elude

>   allude: mention something indirectly
>   elude: escape from someone

among / between

>   among: used with more than two choices
>   between: used with only two choices

amount / number

>   amount: indicates something that can be measured
>   number: indicates something that can be counted

because / since

>   because: for the reason that, by reason of
>   since: from then till now

cite / sight / site

>   cite: (verb) quote
>   sight: (noun) vision, view
>   site: (noun) location

compared with / compared to

>   compared with: used in the analysis of similarities and differences
>   compared to: used to liken one thing to another, as in a simile

complement / compliment; complementary /complimentary

>   complement: (verb) complete or finish something
>   compliment: (verb) praise
>   complementary: (adjective) serving to complete
>   complimentary: (adjective) free of charge

compose / comprise

>   compose: make up or create
>   comprise: consist of, be composed of, be made up of

confuse / complicate / exacerbate

>   confuse: bewilder
>   complicate: make more complex or detailed
>   exacerbate: make worse

decrease / reduce

>   decrease: lessen in number
>   reduce: lessen in amount

dependant / dependent
>dependant: (noun) someone who depends on someone else
>dependent: (adjective) relying on someone or something else

describe / prescribe / proscribe
>describe: give features or details of something
>prescribe: set down as a rule or order
>proscribe: forbid the practice or use of something

determine / estimate
>determine: decide something
>estimate: (verb) compute the value of something; (noun) the value or cost of something

discreet / discrete
>discreet: tactful
>discrete: separate, distinct

do / due
>do: (verb) act or make
>due: (adjective) caused by, scheduled

elicit / illicit
>elicit: (verb) draw out
>illicit: (adjective) illegal, not permitted

e.g. / etc. / et al. / i.e. / viz.
>e.g.: exempli gratia (for example)
>etc.: et cetera (and so forth)
>et al.: et alii (and others)
>i.e.: id est (that is)
>viz.: videlicet (namely)

farther / further
>farther: used with distance
>further: used with time or quantity

fewer / less
>fewer: used with objects that can be counted by number
>less: used with objects that can be measured by amount

good / well
> good: (adjective) of a high standard, of the right kind
> well: (adverb) satisfactorily, in a good way

imply / infer
> imply: suggest or indicate without saying it directly
> infer: form an opinion, deduce from evidence

it's / its
> it's: it is, it has
> its: possessive form of it, belonging to it

lay / lie
> lay: (verb) place, put
> lie: (verb) recline

lead / led
> lead: (noun) a soft metal; (verb) to guide
> led: past tense or past participle of to lead

loose / lose
> loose: not tight
> lose: misplace or have something disappear

many / much
> many: great in number (countable)
> much: great in amount (measurable)

objective / rationale
> objective: goal, aim
> rationale: reason, justification

precede / proceed
> precede: (verb) go before
> proceed: (verb) continue; (noun: proceeds) money, revenue, income

principal / principle
> principal: (adjective) main; (noun) person in charge
> principle: (noun) rule, code

quiet / quite
>quiet: not noisy
>quite: somewhat, rather

regime / regimen
>regime: political organisation
>
>regimen: rules followed

respectfully / respectively
>respectfully: showing politeness
>respectively: with respect to each in the order listed
>[avoid using respectively (see the section "Keep sentences short")]

role / roll
>role: function, part in a play
>roll: (noun) list of members; (verb) move by turning over

stationary / stationery
>stationary: not moving, fixed
>stationery: paper

that / which
>that: used as an essential, or restrictive pronoun
>which: used as a non-essential, or non-restrictive pronoun

their / there / they're
>their: (possessive) belonging to them
>there: in that place
>they're: they are

to / too / two
>to: (preposition) indicating place, purpose, time, etc.
>too: (adverb) also, very, excessively
>two: the numeral 2

unique / unusual
>unique: sole, only, alone, without equal
>unusual: not common

verbal / oral
>verbal: using words
>oral: spoken, not written

while / whereas
>while: during the time that, in an interval of time
>whereas: in contrast to, on the contrary

your / you're
>your: (possessive) belonging to you
>you're: you are

7

WRITING STATISTICS

The two important sections of a scientific paper associated with writing statistics are the Materials and Methods and the Results. In the Materials and Methods section, you should describe statistical methods clearly, accurately and in sufficient detail to allow the reader to know how the experiment was designed, conducted and analysed. In explaining the statistical methods, nothing is as uninformative to the reader as a statement such as, "Data were subjected to analysis of variance using {insert here the name of your favourite statistical software package}". In the Results section, you should provide sufficient statistical information, e.g. P-values, to help the reader interpret your findings, as well as to support your own interpretations. Writing statistics, therefore, is an important part of writing a scientific paper.

Let's assume first that the statistical methods used in your research were correct and appropriate to test your hypothesis. If you are unsure, consult a statistician. Here we will not discuss how to design your experiment correctly or how to analyse your data appropriately. We will discuss, however, how to *write* statistics in the Materials and Methods and the Results sections of your paper. Statistics itself is difficult for many people to

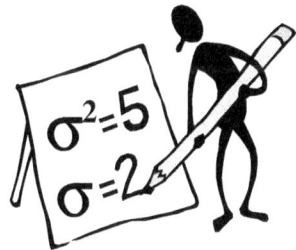

Writing statistics involves attention to many details

understand, so it is important that it is written clearly. To write statistics clearly, the writing should be simple and easy to read, with a clear and well-organized presentation.

MATERIALS AND METHODS

The Materials and Methods section provides the reader with a description not only of the materials and the analytical procedures used, but also of the statistical methods used and the rational for using them if they are unusual. This section describes your material, e.g. location, sexes, ages, diets and numbers. Relevant environmental and experimental conditions, such as climate or nutrient composition, should be described fully. Describe the population or the sample of the population; if there are criteria for inclusion or exclusion, then describe them. Tell how many individuals were included in total, and then tell how many were of each sex, for example, or from each location. It is clearer to write "A total of 100 individuals were studied: 30 males and 20 females from each of two locations A and B," than to write "A total of 100 males and females were used from both locations A and B." With the latter construction, it is impossible to know how the 100 individuals were distributed between sexes and between locations. If age or weight of the individual is important, for example, then give the distribution of ages or of weights. Some journals ask that you use "N" for the total sample (e.g. $N = 100$) and "n" for each sub-sample (e.g. $n_m = 30$ for males and $n_f = 20$ for females). Try to use symbols and subscripts that are suggestive of the factor being described (e.g. subscript m for males).

Tell what type of study you conducted: was it a simulation study, a survey, a field study, an observational (e.g. behavioural) study, or a designed experiment? If it was a designed experiment, then you must have asked yourself the question "What is the experimental unit?". Was the experimental unit an individual or a group of individuals, e.g. a fish or a tank of fish, a chicken or a cage of chickens, a plant or a pot of plants? For animals fed in groups, for example, the group is the experimental unit. It is important to define the experimental unit and to describe how units were chosen or excluded. Were they a random sample of units, or were they a stratified random sample? If stratified by sex or location, for example, what was the rationale for doing so?

If your data were from a designed experiment, in contrast to data from the field, then the design of the experiment should be specified. Describe the design by name and size as, for example, "The study was conducted as a randomised complete block design with six treatment combinations in each of five days (blocks)".

Remember that "cross-classified" ("factorial") and "nested" ("hierarchical") are not experimental designs, but rather they are arrangements of factor or treatment levels in combination with each other. Describe the arrangement as, for example, "The six treatment combinations comprised two levels (1, 2) of Treatment A and three (1, 2, 3) of Treatment B, and were arranged in a 2 × 3 factorial. The control, or reference group, received level 1 of A and level 1 of B".

When describing a design that involves several levels of error, such as a split-plot or a repeated-measures design, be sure to state which error term was used to test which hypothesis.

Describe how your measurements were taken

If you took an initial measurement, took duplicate measurements on the same experimental unit, performed duplicate analyses, or took repeated measurements over time, then describe what you did and how it was treated in the statistical analyses, e.g. as a blocking factor or as a covariate. To describe how measurements were taken, it is clearer and easier to write in the singular rather than in the plural. It is better to write, for example, "Each individual was weighed to the nearest kg", which makes it unambiguous that weighing was by individual, than to write "Individuals were weighed to the nearest kg", which makes it ambiguous whether weighing was by individual or by group.

If you compute a standard statistic, such as the sample mean, the sample standard deviation, or the product-moment correlation, then it is not necessary to give a detailed description or a reference for it. Nor is it necessary to describe in detail or give a reference for a common statistical test, such as the t-test, the F-test, or the chi-square-test. It is sufficient to write, for example, "Goodness of fit between observed and expected numbers was tested by chi-square", without giving a reference for the test. Non-standard statistical procedures or unusual statistical tests, however, should be described clearly and completely (perhaps in an appendix) or be referenced with page numbers; modifications to standard procedures should be described fully.

The statistical model used should be specified; be sure to define the observed dependent variable and the independent variables. Specify also whether an effect is fixed (e.g. location, variety or treatment) or random (e.g. animal, technician or residual). Some journals suggest specifying the model using uppercase letters for fixed effects and lowercase letters for random effects, but practices vary, so consult the instructions for authors. If there are blocks or plots, then describe them. If there is a measured covariable, such as age on a continuous scale, then specify it. If you use orthogonal contrasts to make planned comparisons of treatment means, then define them. If you use orthogonal polynomials to describe the relation among treatment means, then name them. Don't leave it to the reader to guess what you did! It is not useful for the reader if you describe the model only as, for example, "Location, sex and block effects were included in the model". That description raises questions about which effects are fixed or random (if not answered elsewhere), about which is the correct error for tests of hypotheses, about which effects are cross-classified or nested, about the presence of interactions, and about what constitutes a block.

The statistical model might be easier for the reader to understand if you describe it briefly in words before you give the expression. When writing the statistical model, use letters that suggest the meaning of the variables, if possible, e.g. "W" for weight or "L" for location, or simply use words. State clearly any assumptions underlying the model, e.g. that the random errors are assumed to be independent.

An example of a *poor presentation* of a statistical model is:

For both Groups 1 and 2, the usual model is given by:

$$Y_i = a + b \, X_i + e_i$$

to explain the linear relationship between weight and age, where the dependent variable (Y) is a function of the independent variable (X), a and b are parameters, and e is the error term.

Poor presentation of model

First, the presentation lacks motivation; it fails to give the reader a clear understanding of the concept being modelled before giving the model. Second, the presentation is unclear whether the model is being fitted to data for the two groups simultaneously, or for each group separately. Third, it is uninformative as to what is the dependent variable and what is the independent variable (i.e. it is not specified whether Y is weight or age). Fourth, it asserts that the model is usual, but provides no reference to support the assertion. Finally, it fails to define the range of the indicator i, so the number of observations is unknown.

An example of a *better presentation* of a statistical model is:

The relation between body weight (W) and age (A) was assumed to be linear over the course of the experiment and parallel for Groups 1 and 2. The model used to describe the data, therefore, was:

$$W_{ij} = \mu + G_i + bA_{ij} + e_{ij}$$

where W_{ij} is weight (kg) for individual j in group i, μ is the overall mean, G_i is the fixed effect of group i (i = 1, 2), A_{ij} is the age (mo) for individual j (j = 1, ..., n_i) in group i (n_1 = 12 for Group 1, n_2 = 16 for Group 2), b is the pooled within-group regression coefficient to be estimated, and e_{ij} are the random errors, assumed to be independent and $N(0, \sigma_e^2)$.

Before even encountering the model, this presentation states the reason for the model and the assumption of linearity underlying the relationship. The presentation is clear in that the model is fitted to data for both groups simultaneously, and it is informative as to the meaning of the dependent and independent variables. The description of the model defines the range for the indicators i and j, identifies which effects are fixed and which are random, and it gives the assumptions for the random errors.

Good presentation of model

Be careful how you use the word "variable" or "parameter". It is a common mistake to use parameter to describe an observed variable. A variable, is an observed or measured quantity, either discrete or continuous. In contrast, a parameter, in a strict statistical sense, is an unknown constant, such as the mean or standard deviation, which describes the population and is estimated by a statistic. A parameter is often written using a Greek letter, such as "μ" for the mean or "σ" for the standard deviation. For most purposes, however, it is appropriate to use Latin letters for parameters as well; consult the style of your journal to be sure.

There is sometimes a question of where to present results of a preliminary analysis. Do results of a preliminary analysis go into the Results section, merely because they are results, or do they go into the Material and Methods section, simply because they are preliminary to the final analysis? The question arises because you might not want to discuss results of the preliminary analysis in the same detail as you might discuss results of the final analysis. The answer to this question depends on many factors, such as the amount of description of results of the preliminary analysis and the length of the discussion of results. It seems best, however, to present in the Materials and Methods section a brief description of results of the preliminary analysis and a brief discussion of how it led to the final analysis. The results of the final analysis should be presented in the Results section.

RESULTS

The Results section describes your findings, usually with the support of tables and figures. Journals do not want you merely to repeat numbers in the text that are found in a table. Do not write, for example, "Mean of individuals on Treatment A was 12 kg, whereas mean of those on Treatment B was 15 kg (Table 1)". Instead, describe and support your findings with numbers from the table or figure. Write, for example, "Individuals averaged 3 kg less (P < 0.05) on Treatment A (12 kg) than on Treatment B (15 kg) (Table 1)".

Organize the Results section around the tables and figures

It is by convention that we use the term "significant" for P ≤ 0.05, "highly significant" for P ≤ 0.01, and even "very highly significant" for P ≤ 0.001. Use of these expressions is discouraged by some journals. It is redundant to write both "significant" and the probability level, e.g. "Variable 1 was significantly greater (P < 0.05) for Treatment A than for Treatment B". It is better simply to describe the effect that the treatment had on the dependent variable and to give the actual P-value associated with the effect, e.g. "Variable 1 was greater (P < 0.05) for Treatment A (X kg) than for Treatment B (Y kg)". For a regression analysis, write, for example, "For each unit increase in variable X, variable Y increased by 2.3 units (P < 0.05)".

It is misleading to imply that a difference is significant when it is not, by using terms such as "numerically different", "trend", or "tendency". If there are differences or relationships that are "almost significant", but failed to reach significance because of a limited number of observations, it is sufficient to write "Treatments did not differ in their effect on Y (P = 0.056)" and let the reader decide whether the lack of significance is important. You might want to consider reasons for failing to reach significance in the Discussion section, but do not discuss the difference as though it were significant. Do not report P-values to more than three decimal digits. It is no more convincing to the reader to report $P \leq 0.0001$ than it is to report $P \leq 0.001$.

When you report an estimate of a parameter, such as the estimate of a mean or a regression coefficient, also report the measure of precision associated with it, such as its standard error or confidence interval and its degrees of freedom. For a treatment mean, for example, report the standard error of the mean. The standard error (not the standard deviation) is usually what is attached to the mean by "$\pm$", as in mean $\pm$ standard error. This construction, however, might imply an interval around the mean, which it is not. Consider simply using parentheses around the standard error, as in mean (standard error), which allows flexibility in where to place the mean and the standard error in a table; but make it clear that it is the standard error of the mean and not the standard deviation.

In addition to means of main effects, cell means of combinations of the main effects should be given, either in a table, or in the text if there are only a few, so researchers can combine data in the future. For the difference between treatment means, report also the standard error of the difference, even for differences that are not statistically significant. Follow rules for rounding and report only those digits in a number that are meaningful, i.e. that reflect the accuracy of the data. If you weigh individuals to the nearest kg, for example, report the mean as 71.5 kg, not as 71.472 (which implies you weighed them to the nearest gram). Usually it is sufficient to report two or three significant digits (e.g. 123, 3.14, 0.0012), which are not to be confused with decimal places. The number 0.006037 has 4 significant digits (and 6 decimal digits) ... the leading zeros are not significant digits.

WRITING MATHEMATICS

Use as many mathematical expressions as necessary to develop your concept, but do not use more than necessary. It is important for the reader, however, that you show how you developed your expression or, if it already was derived, that you give a reference where it can be found. Place each expression, or group of expressions, on a separate line. If the expression is referred to elsewhere, then be sure to identify it by a number; align the number flush to the right. Consult the journal to see if the number should be enclosed in (parentheses), [square brackets] or {braces}. Keep in the text short expressions that do not require a number. If a mathematical expression appears in the text, in contrast to appearing on a separate line, as in a fraction, separate the numerator and denominator by a slash (/), e.g. 1/2, or use a symbol ½, and avoid text that extends above or below the base-line of the text row. If your expression is displayed in a separate line, as opposed to being in the text, and must be broken because it is too long, then break the expression before an operator (such as = or +) and not after it.

During the revision of your article, remove an expression that is not necessary or add an expression that is necessary, if it will help the reader to follow your development. Remember, however, to renumber each expression and any reference to it. There are usually functions in word-processing and mathematical programs for such updating. It is useful to the reader, sometimes, if you give a numerical example to show how to use your expression, although you probably cannot give an example for each mathematical expression. The numerical example, therefore, should be appropriate and simple. Consult your journal to see if the example should be placed in an appendix.

If you write a mathematical expression in the form of a normal sentence, then do it with proper grammar and punctuation. A mathematical expression (e.g., $W = A$) has a subject (W), a verb (=), and an object of the verb (A). Remember, however, to use the "=" sign only in a mathematical expression, and not in a mathematical sentence; in a sentence, use "is" or "equals". Some journals require a full stop (period), at the end of the mathematical expression. Put a space around a symbol of operation, such as " + " or " / " or " = " (e.g. write $X + Y$ or $4X = 8$), but not when it is part of a fraction (e.g. 1/3). Do not put a space after a function word, such as log or sin, if it is followed by a mathematical argument in parentheses, e.g. $\log(x)$. Negative numbers (e.g. -15) should be written with no space between the negative sign and the number.

Use an appendix for material, such as a mathematical proof or a description of technical details, that would take away from the flow of "telling your story" if it were included in the text.

WRITING NUMBERS, DATES AND TIME

Numbers

If a number is used with a unit of measurement (e.g. 5 kg, 20 days, 14 USD, 75 %, or 30 °C) or with a noun (e.g. Group 15 or Experiment 2), then use numerals. If, however, a number is not used with a unit of measurement or with a noun, then spell out the number if it is less than 10, e.g. "seven treatments", but use numerals if it is 10 or above, e.g. "12 treatments". This is true also when using "-fold" to describe a response: e.g. write fivefold (one word), but 10-fold. Be aware, however, that "-fold" should be used to indicate only an increase and never a decrease.

For numbers less than unity, consult the journal to see if you should use a zero before the decimal point, e.g. 0.25, or if you should write .25. Some journals do not require a comma for a number that is fewer than five digits, e.g. 1000 kg, whereas they might require a comma for a number that is five digits or more, e.g. 10,000 kg. Be aware that there is a difference among countries in the way thousands and decimals are separated. Some countries, such as the UK and U.S., use commas and points (1,234.56) or spaces and points (1 234.56); other countries use points and commas (1.234,56) or spaces and commas (1 234,56). So make sure you use the correct punctuation.

Use "to" instead of a hyphen to indicate a range in numbers, e.g. 30 to 60, not 30 – 60. If it is not clear, you may need to indicate that 60 is included in the range, e.g. "30 to 60 inclusive" in UK English, or "30 through 60" in U.S. English. Use a hyphen for a unit of measure that serves as an adjective, e.g. 100-day period.

For a series of things, some having fewer than 10 and some having 10 or more (e.g two X and 12 Y) use numerals for all things in the series (e.g. 2X and 12Y). If two numbers are adjacent, spell out the first: twelve 2-day-old females, rather than 12 2-day-old females; but write 12 day-old females. Do not begin a sentence with a number; instead, spell out the number (even when it is 10 or more) or reconstruct the sentence. When writing a number that ends with many zeros, use a word for the part that describes the zeros: 16 million rather than 16,000,000. In tables and figures avoid using "$\times 10^{-3}$" or "thousandths" and "$\times 10^3$" or "thousands"; instead use the appropriate metric unit of measurement (e.g. mg or kg instead of g). Do not use more than one slash (/) to indicate "per" in an expression, as in "X g/individual/day", because it is ambiguous as to what divides what. Instead, write "X g/individual per day".

Avoid using numbers or mathematical symbols that can be confused with letters, particularly with certain font types. Especially confusing are the numeral "one" (1) and lowercase letter "el" (l), the numeral "zero" (0) and letter "oh" (O, o), the multiplication sign (×) and lowercase letter "ex" (x), and the upper- and lowercase Greek chi (X, χ) and letter "ex" (X, x). Use a hyphen to join two or more adjectives (called stacked adjectives) or an adjective and a noun that together modify a noun, e.g. goodness-of-fit criteria. It is not necessary to use a hyphen after a prefix such as "re" or "non", unless the word without the hyphen can be confused with another word, e.g. "recreation" (meaning "relaxation") and "re-creation" (meaning "create again"). If you use an abbreviation or an acronym, first spell it out and then give the shortened form within parentheses, e.g. Best Linear Unbiased Prediction (BLUP). If you use matrix notation to write mathematical expressions, use boldface font for vectors and matrices.

Dates

It is easy to confuse dates written only in numerals, because practices vary among countries. For example, 02/10/01 can mean February 10, 2001 or 2 October 2001 or 1 October 2002. To avoid confusion, therefore, it is best to write the date in order of day, month and year, without separating them by commas, but with the day and year in Arabic numerals and the name of the month spelled out. It is acceptable, however, to abbreviate the name of the month (except May, June, and July) when it is used with the day. Common three-letter abbreviations are: Jan, Feb, Mar, Apr, Aug, Sep, Oct, Nov, Dec. This means that you can write 1 Oct 2002. Spell out the name of the month in full if it is used alone or only with the year.

If you refer to a group of years, such as a decade, then refer to the group by adding an "s" without an apostrophe, e.g. 1990s. If you refer to an interval of years, then give the beginning and ending years in full, e.g. 1998-2001, rather than 1998-01, even if the interval does not include a change in the decade or millennium.

Time

To avoid the use of a.m. or p.m. to express time, use a 24-hour clock, e.g., 0:00 (midnight), 08:30, 12:00 (noon), 14:30. As an alternative to using a colon with time, use a full stop (period). If a unit of time is used in the text with a number, then abbreviate it: year or years (yr), month or months

Express time so that it is not misunderstood

(mo), week or weeks (wk), day or days (d), hour or hours (h), minute or minutes (min), and second or seconds (s). If a unit of time is not used with a number, then spell it out. Express length of day in hours and minutes, e.g. 3h 15min, but express photoperiod as a ratio of hours of light (L) to hours of darkness (D), e.g. 16L:8D.

As you can see, a large part of writing statistics, mathematics, numbers, dates and time involves attention to many details. The recommendations given in this chapter are useful when writing in English, but many are applicable also when writing in other languages. Remember that practices may vary among journals, so be sure to read carefully the instructions to authors for the journal to which you will submit your paper.

8

LITERATURE SEARCHING AND REFERENCING

Before writing a scientific paper you must read other papers on your topic, so as to put your research into context. Scientific research is not conducted in isolation – it relates to earlier work, and either supports or disagrees with it. Searching through millions of publications can be a daunting prospect. With a bit of planning, however, it can be made much easier.

Searching the literature should be performed in the same methodical way as conducting the research. You should develop a strategy and follow it in a logical way. An important rule is to record where you searched, which terms you used, and what you found. This saves time, because you do not repeat searches that found nothing, and you can quickly retrace articles that were useful.

SEARCH STRATEGIES

Many people now have access to electronic searches, through either Internet search engines or dedicated bibliographic databases in libraries (see examples in the chapter "Further Reading"). The common feature of all electronic methods is the use of *keywords*.

The author of a paper usually specifies keywords, although the administrators of a database may add words. Sometimes keywords are generated automatically by indexing programmes. Keywords indicate the content of the paper and can be used to search for specific subjects. For example, typing "cat" in a keyword search will list all papers where "cat" was chosen as a keyword. Powerful computers allow faster searching, so some database systems allow you to search the title or even the abstract of a paper for specific words. Do not use keywords such as "the" or "of" – usually they will be ignored.

Keywords normally are combined to form a logical expression. For example, "cat and dog" might be used to find all papers that contained both words "cat" and "dog", whereas "cat or dog" would find all papers that contained either word or both words. Some search engines use "+" instead of "and", and use "/" instead of "or", so it is important to read the instructions. The word "not" or the symbol "-" might be used to exclude words, so the expression "cat not tiger" would list all papers about cats that did not contain the word "tiger". Complicated expressions can usually be generated by using brackets, e.g. "(cat not tiger) and (dog or horse)" would list all papers that contained the word "cat" and either "dog" or "horse", unless the paper contained the word "tiger".

For most search engines, you can use "wildcards" to truncate keywords and widen the scope of the search. The wildcard "?" normally signifies any single letter, and "*" normally signifies any group of letters. For example, "cat?" means search for "cat" plus one more letter, which usually covers the plural form (i.e. cats), whereas "cat*" means search for "cat" with all possible endings (e.g. cats, cattle, catalogue, caterpillar). The "*" should be used with care, and should not be used for words shorter than four letters, because that results in hundreds of keywords and slows search engines.

The secret of efficient searching is to choose your words carefully; if words are too general, you can get thousands of results; if words are too specific, you might miss relevant papers. An example search is shown in Table 8:1. The researcher searched a database for references on "The effect of insulin-like growth factors (IGF) on RNA expression in bovine ovarian tissue". Clearly "RNA" is too general as a keyword because over 100,000 papers were found (Search 1). Adding the keywords "expression" and "ovarian" narrowed the search (Searches 2 and 3). Only when "bovine" was included, however, did the number of papers found become manageable (Search 4). When "IGF" was added as a keyword (Search 5), only 5 papers were identified. At this point, you need to consider whether any of the keywords are too specific. If this is the case, you might miss some valuable references because the author has chosen a different keyword. For example, "IGF" may be written in full (insulin-like growth factors), and expanding the search to "IGF or insulin" adds another six references (Search 6). Similarly, authors may choose "ovary" instead of "ovarian" as a keyword; allowing for this adds a further six references (Search 7). Looking at an index or thesaurus might help you to find alternative keywords for your search.

Table 8:1. Results of a database search showing importance of choosing keywords carefully

Search	Keyword expression	Number of papers found	Comments
1	RNA	101,073	Too many papers to examine individually
2	RNA + expression	85,685	
3	RNA + expression + ovarian	1192	
4	RNA + expression + ovarian + bovine	150	
5	RNA + expression + ovarian + bovine + IGF	5	Too specific, may have missed some papers
6	RNA + expression + ovarian + bovine + (IGF or insulin)	11	
7	RNA + expression + (ovarian or ovary) + bovine + (IGF or insulin)	17	

This example illustrates two principles of searching databases. You can narrow your search by adding more specific keywords; you can widen your search by giving alternative keywords. In this case, we started from a wide search and narrowed it. An alternative strategy would be to start with a specific search and make it wider. There is no best strategy; only you can decide when you have a reasonable number of relevant references.

Electronic databases have revolutionised literature searching, but they do have limitations. Because computers deal with precise facts (ultimately, either 0 or 1), programming them to make approximations or associations is difficult. Language is often imprecise, even scientific language, and there are always at least two ways to say the same thing. The human brain can make associations far more efficiently than any computer. A computer will find a reference only if a keyword chosen by the author or editor matches a keyword chosen by you. Sometimes you might find a reference that is relevant to your topic but contains none of your keywords.

MANUAL SEARCHING

There are specialised publications to help with manual literature searching, such as *abstracting journals* and *contents lists*, although they are being replaced by electronic versions. Abstracting journals gather abstracts from papers published in journals and group them by subject. Looking at a subject area might help you find a useful paper that is described by keywords you had not thought of. Contents lists catalogue the tables of contents from recent issues of selected journals. Again, looking at full titles might help you make an association between your subject area and a particular paper. Keywords can also be used for manual searching.

A manual search might also be needed

Review papers are another valuable source of information, particularly when you are researching a new area. Someone else has already found a list of references for you and interpreted them. Be aware, however, that the author's interpretation might be different from yours, so you must look at the original papers, if only to be sure that spelling and page numbers are correct. Original papers might also refer to previous papers that the reviewer chose to omit. Review papers can be found in many journals and conference proceedings. Doctoral theses are also good starting points for a collection of references. Be careful, however, when using sources that have not been subjected to peer review (e.g. web sites and non-refereed journals).

RECORDING YOUR SEARCH

It is good practice to record your searches to save having to repeat them, and it is essential to record the results so that you know exactly where references come from. Bibliographic software is available for recording references on a computer, and many packages allow you to paste information directly from a database search. Alternatively, you can record the information manually on index cards or pieces of paper. An advantage of using computers for your reference recording is that you can easily search among your references, sort them into the correct order and format them for the reference list of your paper.

You obviously want to record all relevant papers, but it is also important that you record accurately all available information about each paper. For a journal paper you should record the title, all authors' names, year of publication, full and abbreviated journal names, journal volume, part or issue numbers, and first and last page numbers. For papers published in books, you should record accurately the title, all authors' names, title of the book, any editors' names, edition number, publisher's name, date of copyright and place of publication. This information will be needed if you refer to the paper in your own published paper.

You can also record the abstract and keywords for each paper, but you might want to assign your own keywords to the paper and write your own extract of the content. Writing an extract that contains a few key sentences is useful because it forces you to analyse the paper. By doing this, you focus on the important message of the paper and decide what information you might use later in your own review.

Another way to record a paper is to request a reprint from the author or to photocopy it. Photocopying must be performed only within the laws of copyright, which are designed to protect the rights of authors (see the section "Copyright"). Normally you are allowed one photocopy of a paper from a journal for educational or research purposes, but the law varies among countries, so check before you copy. Remember though that photocopying is not a substitute for reading; a good researcher will still analyse a paper and write down the key messages. When the time comes to write your own review, you can rearrange the key sentences from papers you have read, which will provide you with a good starting point.

REFERENCING PUBLISHED WORK

Your work must be related to what other research workers have published. This is done through references that acknowledge your sources of information. If you do not acknowledge your sources, you will be accused of plagiarism, which means representing someone else's work as your own.

The way that references are cited varies among journals, so make sure you read the instructions and look at some papers published in that journal. This lack of standardisation is annoying, but emphasises the need to record full details of each paper in your collection. If a journal requires the first and last page numbers in a reference, for example, and you have only recorded the first page number, then you will waste time by having to find the reference again.

Citing references within the main text

The two major systems of citing references are by *name* or *number*. When names are used, the authors' last names and year of publication are given in the text of your paper, and references are listed at the end in alphabetical order of first author's last name. When numbers are used, only a number is given in the text, and references are given at the end with a corresponding number. These references may be numbered consecutively as they appear in the paper (so the first reference in the text is given the number 1) or they may be sorted alphabetically by first author's last name in the reference list and numbered in that order (so the first reference in the text might not be number 1).

To cite someone's work, using either the name or number system, you make a statement and cite them either after the statement or as part of the statement.

Examples of citing by name or number *after* the statement:

Name: Experimental evidence suggests that the sky is blue (Smith, 1999).

Number: Experimental evidence suggests that the sky is blue[5].

Examples of citing by name or number *as part of* the statement:

Name: Smith (1999) observed that the sky was blue.

Number: Smith observed that the sky was blue[5].

For each journal, you have no choice of system; you have to follow the instructions for authors. Some journals prefer names, because they associate the statements with actual people and dates; some prefer numbers, because they do not interrupt the flow of text.

When a reference cited in the text has more than two authors, you will be asked either to list all authors, e.g. Smith, Brown and Yang (1999), or to use the abbreviation "et al.", e.g. Smith et al. (1999), which is short for the Latin et alii, meaning "and others".

When an author or group has more than one publication in one year, the publications are usually distinguished by adding a letter to the year, e.g. (1990a), (1990b), in both the text and the reference list.

When several references are listed together in the text, they are either in chronological order, e.g. (Jones, 1998; Black, 2000; Green, 2000) or in alphabetical order, e.g. (Black, 2000; Green, 2000; Jones, 1998).

Writing the reference list

The format of reference lists at the end of your paper is different for virtually every journal. For this reason, you should read the instructions for authors carefully and follow them. The only advice that can be given here is to make sure your references are correct and complete. Every reference made in the text should have an entry in the reference list and all references in the list should have been referred to in the paper. An important purpose of the reference list is for others to be able to find the papers you are referring to; incorrect or incomplete references are useless, which is why you need to record accurately all details during literature searches. If you have used bibliographic software to record references (see the section "Recording your search"), you can usually use it to paste references directly into your manuscript. This process ensures that your reference is accurate and complete (assuming you entered it correctly into your reference database). Most reference-managing software allows you to format citations to suit the style of any journal.

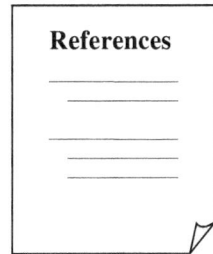

If an author's work has not been published, but you still want to refer to it, do not include it in the list of references. In the text, use a form such as (A.G. Smith, personal communication), where the author has given you some information (and permission to quote it!), or (P.R. Brown, unpublished data), where the information is not yet in print. Again, follow instructions for authors. Note, however, that not all scientific journals allow you to cite unpublished data.

REFERENCING WEB ADDRESSES

The World Wide Web is becoming increasingly important as a source of information. Remember, however, that most web sites are not refereed or checked for accuracy; the contents may change at any time or the web page may disappear without notice. Nevertheless, you are using someone else's work, and you must still acknowledge the source. There are no firm rules, but the following guidelines can be followed in the absence of other instructions:

Within the main text

If the web page is attributed to an author or organisation, cite this in the text using the same system as used for other material, e.g. USDA (1998), and give the full reference in the reference list, as shown below. If the source does not show an author, cite the organisation hosting the web site.

Within the reference list

You should ensure the following information is available:

- The name of any individual or organisation associated with the data/page

- The name of the organisation hosting the site

- The title of the page

- The full web address

- The full date on which you last accessed the site

Example:

USDA. 1998. Agriculture Fact Book 1998, Chapter 1: U.S. Agriculture-Linking Consumers and Producers, Section 1: What Do Americans Eat? http://www.usda.gov/news/pubs/fbook98/ch1a.htm (Accessed 15 Apr 2003).

COPYRIGHT

Copyright exists to protect the rights of authors to benefit from their work. More specifically, copyright is designed to prevent others from benefiting from an author's work without permission, for example, by passing the work off as their own, by avoiding payment when photocopying, or by mounting the work on a web site without acknowledgement. Infringing copyright carries serious penalties, so make sure you always cite your sources of information and seek permission to use illustrations or substantial quotations.

Copyright does not protect an idea; it protects the form or expression of an idea in literary and artistic work. In the case of scientific papers, this means the text, tables and diagrams are protected, but the title is not. The important word is *form*; for example, you are allowed to repeat what someone has written if you change the wording and give a reference. With diagrams, however, you must make significant changes to avoid infringing copyright, unless you obtain permission. You are always safer to obtain permission first, however, because it is usually difficult to know what changes are considered to be "significant".

Copyright is covered by different laws in different countries, but the principles are the same: you cannot reproduce anything in an unmodified form without permission. Protection lasts for 70 years after the author's death, or 95 years from publication of an anonymous work, but it is wise to assume that all printed material is covered by copyright. Copyright is established as soon as original material is created in a form that could be copied. A copyright notice can be placed on the work (e.g. © 2000 Mary Smith) to identify the copyright owner, but this is not a legal requirement. The work is protected even without that notice, but certain legal remedies require that a copy be registered within a specific time period. The copyright notice is often found only on one of the front pages of a book or journal, but it applies to everything in the publication.

Copyright laws generally prohibit photocopying, except that fair use allows a person to make one copy for research or private study. Do not assume that articles found on web sites are free from copyright; the web author either owns the copyright or may have obtained permission to mount the work.

As an author of a scientific paper, you need to be aware of copyright restrictions. The U.S. law is described on the Library of Congress web site (http://www.loc.gov/copyright); the UK law is described on the Copyright Licensing Agency web site (http://www.cla.co.uk), which also has links to sites in many other countries. International rules govern which country's law applies to copyright disputes. In most copyright laws, however, there is reference to "fair dealing" or "fair use", which allows you to reproduce material (e.g. photocopying as above) for research or private study, or for purposes of criticism and review, without obtaining permission. The laws do not state how much you can reproduce in a review, but passages of up to 400 words from a book or 5% of an article are often accepted as upper limits.

Only the Courts and not the author can decide whether you break copyright limits, so it is best to seek permission from the copyright holder if you use more than one or two sentences. Permission is not needed if you prepare a table or figure from data published elsewhere, for example, by combining data from several papers or by drawing a graph of data published in a table. In each case, however, you should acknowledge the source(s) with a proper scientific reference (see the section "Referencing Published Work").

The copyright holder is usually the author of the work, but copyright can be transferred or a licence can be granted to the publisher; this will be shown in a copyright declaration. If an author produces a literary or artistic work as part of employment, e.g. a journalist or writer of instruction manuals, copyright usually belongs to the employer. If you want to reproduce substantial text, a table or a diagram from someone else's paper, ask for permission. You normally should write to the publisher stating what you want to reproduce (full reference), where you want to reproduce it (full provisional reference), and how you intend to modify the material (if you want to change a diagram, for instance, enclose a copy of your proposed version). The publisher may charge a small fee for permission to reproduce the material; if the fee is too large, either seek advice from your own publisher or omit the material. A sample permission form is:

> Dear Sir or Madam,
>
> I am writing a [paper / chapter / book] provisionally entitled ..., which will [be submitted to the journal ... / be published by ...]. I would like permission to reproduce the following material from the paper [*Full reference*].
>
> [Figure X, Table Y]
>
> [I would like to change the regression line in Figure X to a solid line because I feel this will be clearer than the current dotted line].
>
> Full acknowledgement will be given to the source of the material; if you require a particular wording, please let me know.
>
> Thank you for your assistance. I look forward to hearing from you.
>
> Yours faithfully,

Larger publishers have permissions departments, so address your letter there. If the publisher refuses permission, seek advice from your own publisher. You may be able to modify the material sufficiently to be outside the original author's copyright. Under no circumstances, however, should you use the original material without permission.

9

GETTING A PAPER INTO PRINT

Research that yields new knowledge should be written up as a scientific paper and submitted for publication to an appropriate peer-reviewed journal. The peer-review process is essential to maintain scientific standards and to ensure that your paper is worthy of publication. In addition to reviewers, editors check the quality of scientific writing and clarity in presentation of concepts. Submission of the paper to more than one journal is considered to be a breach of professional ethics. A journal will not knowingly accept a paper that has been, or is scheduled to be, published elsewhere; an abstract that was published for a conference is an exception.

The choice of journal depends on who you want your audience to be … who do you want to read your paper? You may choose a journal where other papers in your field are being published; you should know which journals are relevant from your review of literature. Your choice may be limited if your paper is specialised. If there are several options, go for a high quality journal that is widely read. Visibility, in itself, is important because publishing in a journal that few people read does not help disseminate your research.

PREPARING YOUR MANUSCRIPT FOR SUBMISSION

Before you submit your paper, consult the journal for the details (e.g. the number of copies), the form of submission (e.g. paper or electronic), and where to submit it. Make sure you follow formatting instructions about page layout (margins etc.), line spacing (usually double), line numbering (usually required), referencing, tables and figures, visuals, and other style issues before you start writing. Although journals will edit accepted papers for style, clarity of writing, and correct grammar, they might reject your paper if you have not followed their instructions.

Some journals require a non-refundable processing fee for each paper submitted. Many journals ask authors to complete and submit a "Manuscript Submission and Copyright Release Form", which assures that each author has read the paper and that the paper is not being submitted for publication elsewhere, and which releases copyright of the paper to the journal. Copyright is discussed in the chapter "Literature Searching and Referencing". You do not *have* to release copyright to a journal, but you may grant the journal a licence to use your material. In some instances you may not be *allowed* to release copyright – either because it is not yours to release or because it belongs to the public, e.g. if you are a government employee.

AUTHORSHIP AND ADDRESSES

There is great debate about authorship of papers, including who should be listed as an author and in what order. These matters ideally should be agreed at the start of the research programme, but decisions are often postponed until the paper is nearing completion, so we are highlighting potential problems here.

The list of authors should include only those who made a substantial *intellectual* contribution to the research work, to the writing of the paper, or to both. Do not include people in the list unless they were actively involved in the project; the practice of putting the head's name on every paper is no longer valid. Similarly, do not put a colleague's name on a paper in return for getting your name on his or her paper; this dilutes the credit given to the other researchers. An ethical scientist will not ask to have his or her name included, nor include anyone else's name, unless the person has been involved closely with the work. It is important, however, to make sure that everyone who has made a substantial contribution *does* get listed as an author. The contribution should be intellectual, e.g. formulating the original concept, obtaining the funding through a grant application, designing the experiment, supervising the project, or writing the paper. Lesser contributions should be listed in the acknowledgements, e.g. a technician who collected the data, if the person was merely following instructions. All authors jointly share credit and responsibility for the contents of a paper, so make sure that all authors have read the final version before submission.

The decision on order of authors is easy when only two people have contributed to the work; the person who does the research and writes the paper usually comes first (e.g. first the student, then the supervisor); if the other person substantially rewrote the paper, the order might be reversed. The decision on order of authors is more difficult, however, when the paper reports work from a team. Research groups have different conventions: alphabetical order,

order of contribution (greatest contribution first) or reverse order of seniority (Professor last, although note that the first author is sometimes referred to as the "Senior Author"). If the team is from more than one institution, authors may be grouped by institution. The order is immaterial, except that the first author of a long list will have the honour of being referred to by name, with other authors referred to "et al.". One author might like to be the last author on a list, because it implies seniority in some systems, but this is offset by other systems where the last author made the least contribution. There is no single way to order names on a paper; it is different for different groups ... it depends on the culture of the group. To avoid misunderstandings and potential conflict, it is best to have open communications among possible authors early in the research process about who will be an author on the paper and in what order.

Addresses of authors should be given in the journal style. Make sure the full international address is given, so that interested people can contact the authors. The first address should be the institution where the majority of the work was conducted (normally that of the first author). If the first author has relocated after completing the work, the address should remain the original institution. With more than one institution, superscript symbols are often used to indicate who works where; consult the journal as to the use of symbols and their order. One author may be identified for correspondence and reprint requests.

SUBMISSION

Many journals now encourage electronic submission of manuscripts, but you might have to use traditional postal services. Double-check that you have followed all the instructions for submission and send your manuscript to the address shown in the journal. In an accompanying e-mail or letter, state your name and full contact details (postal address, including ZIP or postal code and country; telephone; fax and e-mail address), together with the title and authors of your paper. You can include a brief statement of the objectives,

You will have to wait a few weeks for the review

which might help the editor choose a referee; some journals encourage authors to suggest a referee. If posting your manuscript, then use a strong envelope or package to avoid damage in transit. You should receive an acknowledgement of receipt within a few days. You will then have to wait a few weeks to see the report of the review. During this time, your manuscript will be considered carefully by an editor and usually two referees; this is the peer review process (see the chapter "Reviewing Papers and Presentations").

EDITOR'S AND REFEREES' REPORTS

Eventually you will receive reports from the editor and the referees. Each referee's report will give opinions on your paper and normally will make suggestions for improvement. Often the editor's report consists of a covering letter that interprets the report from the referee(s) and gives a decision about whether your paper is acceptable for publication.

Most journals have four possible decisions:

1. Accept as written

2. Accept with minor revision

3. Accept with major revision

4. Reject

Decision 1 (Accept as written) is unusual (less than 5%), particularly for inexperienced authors. Most referees and editors will recommend at least some small changes.

Decision 2 (Accept with minor revision) is normal for a paper that describes sound science and communicates that science well. You can usually deal with minor changes quickly and return the manuscript to the editor for publication. Make sure you read the recommendations carefully and follow them if they are acceptable. Do not be afraid, however, to question any recommendation with which you do not agree. Referees and editors can make mistakes or may not understand something you write. If this is the case, however, then you should still consider rewriting the relevant section so that your meaning is clear and unambiguous. Some journals require a covering letter with the revised manuscript detailing how you have addressed each point in the referee's report, especially those points where you disagree.

Decision 3 (Accept with major revision) is similar to *Decision 2*, except the changes requested will be either more involved, more numerous, or both.

Again, read through the recommendations carefully and see if you agree with them. Suggested changes usually will improve your paper, and you should follow them carefully. If there are any that you do not understand, ask the editor for clarification. Editors welcome such requests, especially if it saves them having to return the paper to you at a later stage. Always make your queries clear and short. You do not need to repeat large sections of the paper or the report, because the editor will have copies. Examples of major revisions include recalculating data, correcting statistical analysis, adding more detail to materials and methods, and lengthening or shortening the discussion.

Decision 4 (Reject) can be a shock. You have spent a long time gathering data and preparing a paper, only to have it rejected by anonymous referees to whom you cannot complain. You may feel angry or disheartened, both of which are natural feelings. When you have recovered, however, examine the reasons for rejection. Has the referee misunderstood something in your paper? Can you change the emphasis of the paper to make it acceptable? Can the decision be reversed if you *resubmit* the paper after major revision? Is the material more appropriate for a different journal? If you have a valid reason for appeal, then write to the editor, but first check your arguments with an experienced colleague. If the paper was rejected because of flaws in experimental design or low scientific merit (assuming you described the experiment accurately), then there is nothing that can be done to recover the situation. If the reviewers want you to repeat measurements because you lacked the necessary control, for example, but the experimental material has been disposed of long ago, then accept the decision gracefully and learn from your mistakes. Do not waste other people's time by submitting the paper to another journal. If a less discerning journal does accept your paper with fundamental flaws, then it will not enhance your reputation, and you will be embarrassed to read it in later years.

AUTHOR PROOFS

Once accepted by the editor, your paper is forwarded to a technical editor to prepare it for publication. The technical editor looks in detail at the way your manuscript is written, rather than at its scientific merit. The technical editor may contact you for missing information (e.g. incomplete references), but normally the paper is typeset, figures are reproduced, and author proofs are prepared.

Author proofs are sent to the corresponding author indicated on the title page of the paper. Journals that use electronic handling of papers usually e-mail proofs as a non-editable pdf file, which can be printed out, corrected, and faxed back to the journal. To make corrections to the proof for English-language journals, standard proof correction marks are available, as published in BS 5261C[1]. These standard marks were incorporated into the international standard ISO-5776 and were endorsed by U.S.A., Russia, China and most European countries. Standard proof correction marks can be obtained from the journal and may accompany the proofs, or they can be downloaded from http://www.ideography.co.uk/proof/marks.html. Some journals in the U.S. use a different set of proof correction marks, which can be found at http://www.indiana.edu/~iupubs/proofmarks.html. Author proofs must be dealt with promptly, so as not to delay publication. You will probably be reminded that only typesetting (formatting) and factual errors should be corrected at this stage; if you try to introduce new material you may be charged.

Before you start reading author proofs, make a copy and work on the copy, rather than on the original. Mistakes in your proof correction marks can then be changed without causing confusion for the typesetter. Firstly, read your proof through quickly or, even better, ask someone else to read it, while you follow the manuscript. This will help you find any words that have been left out or any new words that have been inserted. Secondly, go through the proof carefully, checking the spelling. Professional proofreaders sometimes start at the end and read the proof backwards to discover the errors more easily. Finally, check every number in your proof, especially those in tables, against the original manuscript or, if possible, against the original data. You are the only person who can verify the data; incorrect numbers will still look "correct" to other people. When you are satisfied that you have identified all the corrections, enter the proof marks onto the original proofs, and write your corrections neatly and clearly in the margins of the proof, following the proof correction marks. Return your corrected proof, together with the manuscript, to the technical editor.

An example of a corrected proof is shown in Figure 9:1 (proofs normally do not contain so many errors):

---

[1] BS 5261C : 1976. Marks for Copy Preparation and Proof Correction. British Standards Institution, 2 Park Street, London W1A 2BS.

The ABC of Science Communication

Communicating science usually means communicating <u>new</u> knowledge or summarising the present state of knowledge. It is important for the audience to catch the message with as little misunderstanding as possible and to feel confidence in what is written or said.

italics

The ABC of science communication is that it should be:

- Accurate and Audience adapted
- Brief
- Clear.

Science is international. This means that most of those who read/listen to a scientific presentation will be doing so in a foreign language. This further emphasizes the need for clarity and for the presentation to be logical, consistent and coherent. Communication is a two-way process. Information cannot merely be delivered - it must be well received and understood as well. The message delivered may be accurate, brief and clear, but yet not be received and understood. This may happen if what you write or say does not relate to the frames of reference of audience. Adapting to the audience, therefore is very important.

many / or
s
the

A basis for the process scientific is to formulate a hypothesis, which means that you pose a question and a hypothetical ANSWER. Questions and answers are the basis for communication as well. For effective communication you cannot just think of your own topic and the message you want to deliver. You must also consider what questions your audience might have with regard to your topic. Some components of effective communication are indicated in Figure 1:1.

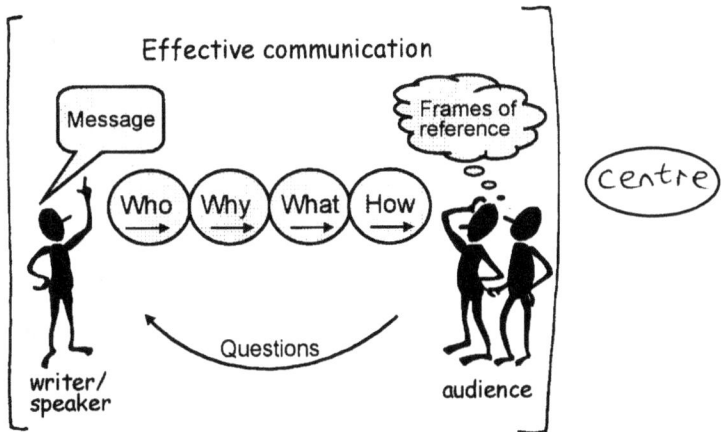

l.c.
stet.

centre

Effective communication

Message

Frames of reference

Who Why What How

Questions

writer/ speaker

audience

Figure 1:1. Some components of effective communication.

Figure 9:1. Example of a corrected proof.

Table 9:1 shows examples of common proof marks that may be used to correct a manuscript. If you do not feel comfortable using these marks, underline or circle the text and write your correction in the margin, clearly, unambiguously, and in simple language. If you write instructions in the margin, draw a circle around them so they are not confused with text to be inserted or changed. If you write "align" in the margin, for example, then you might mean "align columns of text", or you might mean "insert the word align". Your corresponding mark in the text should make your intention clear, but if "align" is circled it is obviously an instruction. You could make instructions even more distinctive by writing them in a different colour to the one you used for text to be inserted or changed.

Table 9:1. Example proof marks (based mostly on BS 5261C:1976)

Mark in text	Mark in margin	Meaning
⋏	*a*⋏	Insert character (*a*)
⋏	*new*⋏	Insert word (*new*)
⋏	⟨A⟩⋏	Insert extra text provided on a separate sheet marked A
/	∂	Delete character
⊢⊣	∂	Delete word(s)
⫯	∂̑	Delete character and close up
⊜	∂̑	Delete word(s) and close up
/	Correct character	Replace character
⊢⊣	Correct word	Replace word(s)
⌣	⌣	Close up (less space)
Y	Y	Insert space

Table 9:1 Continued

Mark in text	Mark in margin	Meaning
‖	*align*	Align columns
≡	*align*	Align characters in a row
⌐	*new para*	Start new paragraph
⌐	*run on*	No new paragraph
⌐⌐	⌐	Swap characters or words
2 3 1	1 2 3	Change order of words
⊢⊏ ⅃	⅃	Move words left
⅃ ⊐→	⅃	Move words right
⊏⊐	*move*	Move words to given position
[ ]	*centre*	Centre
⅄ or /		*Insert or delete:*
	⊙	Full stop or decimal point
	⊙ ⊙	Comma, semi-colon
	ᵞ	Quotation marks
	⊘	Slash
____	*italics*	Change to italics
∿∿∿	*bold*	Change to bold
____	*l.c.*	Change to lower case
____	*CAPS*	Change to upper case (capitals)
	/	Mark between corrections in same line
------------	*stet.*	Leave unchanged (mistake in proof-reading)

If you have two or more corrections for one line of text, separate each correction in the margin by a slash (/). You can use all four margins, but the order must be from left to right across the side margins, in the same order as the line of text. If there is not enough room for a correction in the side margins, place it in the top or bottom margins (clearly indicated), or on a separate sheet of paper. Never write on the back of the proof or your corrections might be missed. Do not enter corrections between lines of text because such corrections might be hard to read. Do not use capital letters when marking corrections (unless the material should be in capitals); the typesetter might set the corrections in capitals.

Some journals charge authors for each printed page, to help cover the cost of publication. A form for ordering reprints is forwarded to the corresponding author with the author proof. The form normally indicates the number of pages for your paper, the charge per page, and the charge for reprints; make sure you return this form with your proofs.

10

ORAL PRESENTATION AND
VISUAL DISPLAYS

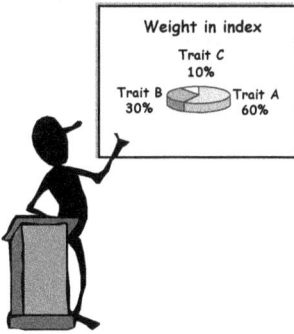

New research results, as well as updated reviews of specific research areas, are communicated at scientific meetings, such as national and international conferences, often before the results are even published in scientific journals. This communication usually occurs in the form of a conference paper (sometimes just an abstract), and an oral presentation or a poster. Oral presentations are also given frequently at other types of meetings, for instance at workshops and university seminars. We will focus mainly on oral presentation at scientific meetings and seminars, but the information is valid also for other types of presentation.

Many different styles of presentation can be successful, as long as your message is conveyed effectively. In fact, variability among presenters adds to the quality of a meeting. Choose a form that fits your personality, but make sure that you adapt to the audience, that you speak to be heard, that you use visuals that can be seen, and that you show an interest in the topic!

PLANNING THE ORAL PRESENTATION

It is worthwhile making an effort to perform a good oral presentation, not only because audiences usually remember both the good presentations and the bad ones, but also because the message is conveyed most efficiently in the good ones. Considering the amount of work that goes into your research, you should make the most of every opportunity to communicate the results and conclusions, to spread knowledge and stimulate action, and to make valuable contacts.

You have a responsibility to your audience; their time is precious. A presentation should add value to merely reading your paper: it should arouse interest in the topic and stimulate discussion, and it should make it easy for the audience to understand the main points quickly.

Who? – Why? – What? – How?

The reason for your scientific presentation at a meeting is usually that you have new research results to communicate, or that you have been asked to review a topic. What you present depends on the purpose of your presentation, what message is interesting and important to communicate, and who your audience will be (see the chapter "Communicating Science"). Adapting to the audience is even more important in oral communication than in written communication. The listeners have to understand what you say immediately; they cannot go back as they can in written text. So, think ahead about your audience: what are their backgrounds, interests and capabilities? Your speech will not be understood and retained unless it relates to the frames of reference of the audience. The audience is often quite diverse with regard to their knowledge of your topic. Your oral presentation, therefore, needs to be at a more general level than your written paper. At the same time, however, you shouldn't lose in-depth coverage of essential parts.

It is sometimes pointed out that the word "presentation" contains the word "present". You are giving your audience a present, something you like to give and hope they will enjoy. Your gift has value, which is enhanced by you being present at delivery.

Some features of scientific meetings

International meetings, where many participants speak and listen to a foreign language, are different from national meetings, where most participants speak and listen to their native language. Some meetings concentrate on a specific topic, whereas others cover a larger area, often split into a number of parallel sessions. In those meetings, the audience is often mixed with regard to research specialisation, and not everyone has a specific interest in every presentation. Usually written papers or summaries are available at the presentation, but the audience often does not read them in advance. Sometimes only an abstract is provided.

The audience at meetings is often large, and questions and discussion generally take place at the end of each presentation or each session. The time schedule is tight, because there are normally a large number of oral presentations in a day. The presenter of an invited paper might get 20 to 30 minutes, whereas the presenter of a research paper is often restricted to 10 to 12 minutes. This restriction further emphasises the need for oral presentations to be attractive, interesting, easy to understand, and focused on the main message. Listening to presentations for many hours, often in low lighting conditions and in a foreign language, is hard for the audience.

PREPARING THE ORAL PRESENTATION

When preparing an oral presentation, you need to think of how best to raise and maintain the audience's interest, and how to structure the topic so the audience will understand and remember your main messages.

Audience attention varies during a presentation

As soon as a presentation stops being interesting, the thoughts of your listeners will easily slip away to something else. For a message to be absorbed, you, the speaker, must be more interesting than the listener's own thoughts. Disturbing factors in the room, e.g. noise from the projector or from people moving between sessions, can also cause parts of your presentation to be missed by the audience. Audience attention varies during a presentation: it is at its highest level at the beginning and at the end (Figure 10:1).

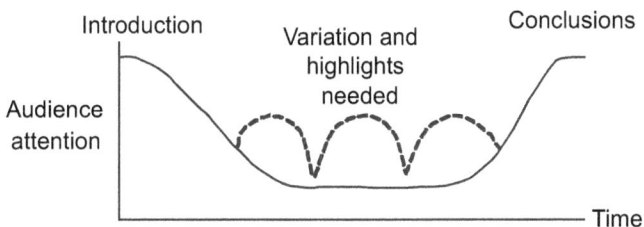

Figure 10:1. Audience attention varies during a presentation.

Audience attention is high at the beginning of your presentation, because the audience is curious about what will come; attention is high again at the end, in anticipation of your conclusions. Attention is lowest, however, in the middle of your talk, so variation and highlights are needed to maintain interest. Because the audience can easily miss or mishear parts of a presentation, the most important points should be emphasised twice to give the listeners a reasonable chance to understand them. Repetition, if possible with some variety, increases the transfer of your message. Plan your talk so that you use the times when audience attention is at the highest level to get across your main "take-home" messages.

Anticipate audience questions

You should consider not only what you want to achieve with your oral presentation, but also the needs of your audience. One way to do this is to write down a number of questions that you expect your audience might have with regard to your topic, and incorporate your answers to these questions into your presentation.

Anticipating audience questions will:

- Motivate you to prepare a good presentation

- Make your presentation better adapted to the audience

- Be helpful in the discussion following your presentation

Many people in the audience probably will be more interested in the importance and possible implications of your results than in the research per se. Generalists will want to know how the results apply to them, whereas specialists may be more interested in the methods used. If the audience is mixed, you should do your best to satisfy their needs, but don't go into too much detail. You might invite those specifically interested in details to discuss them with you individually later on.

Make a presentation outline

Before developing a presentation outline, you must know how much time has been allotted for your talk. Most of us overestimate the amount of information that can be presented in a fixed time period. It will save work later, therefore, if you plan your timing from the outset of your preparation.

It is also wise to consider the language for your presentation. Will it be in a language other than your mother tongue? In that case, prepare your presentation directly in the foreign language, if possible. Doing so will make it easier to start thinking in that language, and you won't get stuck with words or sentences in your native language that may be difficult to translate. If you are fortunate enough to make your presentation in your mother tongue, remember that for many in the audience *your* native language might not be *their* native language, so plan your presentation to be simple and clear.

There is no single way to outline a presentation; that depends on the subject, the audience, and your personality. The same structuring principles that were discussed for written communications, however, can be applied to oral presentations (see the chapter "Getting Started in Writing"). You might choose a structure with regard to chronological events, interest and importance, cause and effect, or "pros and cons".

The chronological structure resembles the scientific writing IMRAD model. Using that for a presentation probably means starting with the historical background and objectives, then presenting materials and methods, results, discussion and conclusions. This model very much relates to your research work, although it might not fully relate to the needs of the audience.

General structure for a presentation

You may not want to follow strictly a single structuring principle; instead you may prefer to synthesise several structures. A general structure for a presentation outline could be:

- Title

- Introduction

- Body

- Summary and Conclusions

Title. Make a good title for your presentation and the corresponding paper. The title will be given in the meeting programme, and should encourage people to come to your presentation and to read your paper. Try to make the title both informative and inviting. It should be as direct as possible, and not too long. Don't be afraid to pose a question, or even to make the title provocative;

that usually attracts the audience. Make sure the audience knows the title before you start your presentation; otherwise they might be "lost".

Introduction. Take advantage of the audience's higher level of attention at the beginning of your presentation to make them interested in listening to the rest. Give the context, background or motivation for the topic you will present. Make the audience understand that your message is important to them. One idea is to start with some of the questions you predicted that the audience might have. Another idea is to tell the audience what they can expect to learn, or what they will go away with. It is recommended that you briefly outline your presentation plan. Be careful, though, not to do this in too much detail, or your listeners might lose their desire to listen further. The introduction must not be too long.

Body. The body constitutes the major part of your presentation. Audience attention may drop during this part, so it is important to structure the content to be both logical and interesting, and to relate to the audience. Use analogies, if helpful, to explain things better and to answer some of the questions you formulated during your preparation. Interpretation and analysis of facts, comments upon problems and prospects, and advantages and disadvantages of various solutions are all vital aspects to include in the body. If you are making a theoretical presentation, try to give realistic examples.

Divide the body into sub-sections, and structure each in the best way. If the audience is mixed with regard to specialisation, then it is probably better to give only a brief outline of the methods and focus mainly on the results. If you apply the chronological model, then the methods will be presented before the results, whereas if you apply the interest-importance approach then the results will come first.

To maintain audience interest during the body of your presentation, it is essential to build variation into it. Using visual aids, body language and voice modulation may all help. Another option might be to allow, and even to encourage, questions from the audience during your presentation. This may be difficult to handle in scientific meetings, where the audience is often large, but it may be well suited for workshops and seminars.

Summary and Conclusions. This is the most important part of the presentation, because what you say here is what the audience will normally remember most. Use this opportunity, therefore, to get across your main "take-home" messages. Summarise your main findings and conclusions for their interpretation and significance or impact, and give your recommendations.

Make sure you get to the Summary and Conclusions part of your presentation before your time runs out. Remember, however, that audience attention will drop again if this part is too long. Limit the number of statements, therefore, and keep them concise and to the point. Plan an "exit strategy" to your presentation; otherwise you might be left standing in front of your audience wondering what to do next. You might end your presentation with a few words to indicate that it is completed, and, if applicable, to stimulate questions and discussion. Remember to acknowledge your co-workers at the start or end of your presentation.

VISUALS SUPPORT YOUR TALK

Supporting a talk with visuals is useful in communicating science, as well as in providing variation and in stimulating interest throughout the presentation. People absorb more information, and retain that information longer, when it is communicated to them in both words and pictures, compared with words or pictures alone (Figure 10:2).

It is essential that the visuals used are seen clearly and are meaningful to the audience. They should relate to the words spoken, be well organised, and emphasise the important points. A visual that is overloaded, that is difficult to read or understand, or that is not apparently related to your talk will only be distracting.

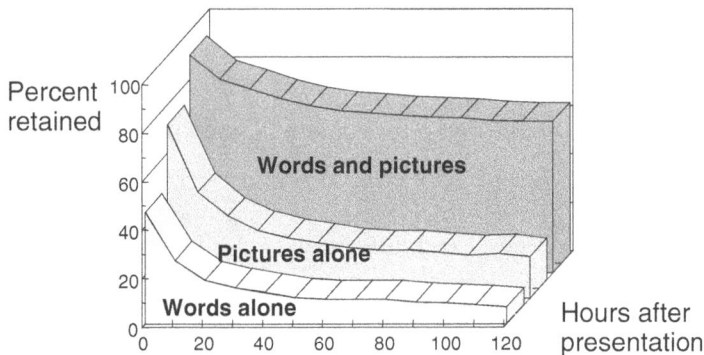

Figure 10:2. Audience retention of presented information[1].

---

[1] Modified from Woelfle, R.M. (Ed.): A guide for better technical presentations, p.37. IEEE Press, New York (1975).

Keep room lights on as much as possible

Lighting of the presentation room is an important factor to consider when preparing visuals. If the room light is on during your presentation, you can glance at your notes without being tied to a lectern. It will also be easier for you to keep eye contact and to "feel" the response from the audience.

The audience will benefit if the room is well lit during your presentation. They will:

- See you better and experience more eye contact

- See to make notes

- Not get tired so easily

Use colours in your visuals that minimise the need to reduce the lighting of the room. If the light has to be reduced, tell the person handling it not to darken the room more than necessary, and to turn on full light again as soon as you no longer use your visuals. The room light can usually be fully on when transparencies are shown. The light can also be on during electronic presentations, assuming powerful modern equipment is available and you use slides with a light background. If slides for a slide projector are used, however, the room light might need to be reduced or switched off, especially when slides have a dark background.

Which visuals to use?

The visuals mainly used in oral presentations at scientific meetings are:

- Electronic presentation
- Overhead transparencies (acetate sheets)
- Slides for a slide projector

Other visual options include films and videos, and in a small meeting flip charts or a writing board.

To decide which type of visuals to use, you need to consider both what technical equipment will be available in the presentation room, and what equipment is available to produce your visuals. Remember that using the most modern methods to produce and show visuals is no guarantee of a successful oral presentation. In fact, technical sophistication might even be a disadvantage, because there is more that can go wrong, and it might require practice and good luck. You can do a lot with a little. If you don't have access to a colour printer, for example, you can still add colour to transparencies or make a drawing by hand.

Electronic presentation

Your visuals can be shown effectively by using presentation software (such as PowerPoint®) and a projector. This method has many advantages and it is used more and more. Before planning an electronic presentation, however, make sure that the room has the necessary equipment, and that the equipment is compatible with your version of software and computer disks. Moreover, learn how to solve compatibility problems, especially problems with resolution.

An electronic presentation offers several possibilities, such as:

- Smooth transition from one display to another

- Sequential build up of bullet points

- Animation of tables, figures and other illustrations

- Displays hidden in readiness if needed

- Flexibility in where to stand, if a remote control is used

Presentation software provides you with many options for transition and animation effects, but it is best to use them with discretion. Remember that the focus should be on the content of your presentation. The effects used should help to emphasise your message, not to distract from it.

Sequential build-up is an option that can be useful. You can reveal bullet points one at a time, and you don't need to use a pointer. Previous bullets in the same display can be set to fade slightly, or dim, but it is best not to make them disappear. You can also present a line diagram or a flow chart sequentially, and you can highlight a number in a table, for example, while discussing it in your talk. Remember that the message might be better conveyed if a bullet point just "appears" on the screen, rather than coming in "flying" or "crawling" etc. In a flow chart or a line diagram, however, the message might be emphasised if sub-parts "wipe" in, for example, in a logical direction. Avoid having too much animation within each display; if necessary, group items to reduce the number of steps. It can be difficult for the audience to grasp the meaning of the text, table or figure if it is changing all the time.

With electronic presentations, it is normally possible to keep room lights on, assuming you choose colours for text and background that give the best conditions for this (see the section "Choose colours and background ..."). Be aware, however, that colours might look different when projected, compared with how they looked on your computer screen.

There are things that might go wrong in an electronic presentation, so it is wise to be prepared with back-up transparencies, especially if your presentation will be abroad at an international conference. It is also wise to bring an extra copy of the presentation file on one or two external disks (floppy, USB, CD, zip disk). Having a printout of your presentation can be wise as well.

Some problems might occur when your presentation is done using a computer other than the one you used to produce your visuals. Scientific symbols, such as μ, might appear as something else if, for example, the presentation computer has a different version of the software or operating system than the one you used at home. The same problem might occur for bullets. One way to reduce such problems is to use fonts that have been used widely for many years, e.g. Symbol, Arial, Verdana, Times New Roman. Other options are to use an equation editor for scientific symbols, and to use pictures as bullet symbols. If you use PowerPoint®, you can save your presentation with "Pack and Go", which packs all the necessary fonts and files used in the presentation together on a disk.

For some advice on showing visuals in an electronic presentation, see the section "Performing the oral presentation".

Overhead transparencies

Using an overhead projector and transparencies is usually reliable, and the room light can normally be on fully. You usually can stand next to the overhead projector to switch transparencies, which might help to make your presentation alive and varied, compared with just standing behind a lectern. You can also reorganize, add or delete

transparencies late in your preparation, or even during the presentation, without the audience noticing.

One disadvantage in using transparencies is that the overhead projector is not always placed where it is most convenient for the speaker. The projector is sometimes not even placed on the stage, and you might then need someone to assist with your transparencies. Organisers should place the overhead projector where it is convenient for you, the speaker, and neither the projector nor the speaker should stand in the way of the viewers. There should be a surface nearby to place your transparencies, and a spare projector bulb should be available. Moreover, the projection screen should be at the correct angle, so as not to distort the image, and be large enough for the size of the room.

Transparencies are usually brighter and better focused if not viewed through a plastic folder. The string of holes is also disturbing, unless you hide it. If you intend to take the transparency out of its folder, do so before your presentation. Speakers sometimes place a paper between each transparency, but you should be aware that this may cause both trouble and noise. For further advice on handling transparencies in a presentation, see the section "Performing the oral presentation".

Slides for a slide projector

Slide projectors have been used widely in presentations. They are now being replaced by electronic presentations, but where equipment for an electronic presentation is not available, slides might still be used. Slides can be used to show text, tables, figures, photos and other illustrations.

Using a slide projector in a presentation might have some disadvantages. Room lighting, for example, usually needs to be reduced when slides are shown, and the speaker gets tied to the lectern, where the only light is. Slide projectors are sometimes noisy, which might be a disturbing factor. Furthermore, you cannot easily exclude a slide from your presentation without the audience noticing, and it may be difficult to go back and show a specific slide again in the discussion.

For the preparation of slides, be aware that it may take several days to get them produced from your computer displays. Be aware also that the standard thickness for slide-frames varies. A European standard frame (3 mm) might get stuck in a projector magazine in the U.S., where the standard is 2 mm. Frames thinner than 2 mm, e.g. paper frames, sometimes fall straight through and block the projector. Find out, therefore, what standard to use before you frame your slides!

CREATING VISUALS

All the types of visuals discussed above, i.e. electronic presentations, overhead transparencies and slides, can be created on a computer, and the same display can be used for all three purposes. To produce the visuals, you can use software for presentation, text processing, graphics and drawings. You can insert clip-art and photos into your visual displays, as well as visual information from other sources, e.g. the Internet. Make sure, however, that you are not infringing copyright.

Choose colours and background that emphasise the message

Visual displays can be produced on a computer in various ways. You can use autolayouts and design templates that help you build displays in a standard format, or you can start from a blank page and give your visuals a more personal design.

Applying a design template harmonises the visuals within your presentation, so they will all have the same background colour and pattern, such as a spiral, ribbons or spots. Having the same background appearance for all visuals in a presentation might seem professional, but it also can be boring, especially if the previous speaker used the same template as you. Showing a visual that does not have exactly the same style as the previous one can be an effective way of stimulating the audience, but you should still strive to be consistent in the use of font(s), font sizes and colours within your presentation. Remember that the background used in displays must not be distracting; if it is, you would be better off with a plain background.

Colour can make visuals more attractive, but using too many colours in a single display can distract the viewer from your message. Remember to:

- Make a good contrast between text and background, i.e. dark text on a light background, or vice versa

- Use contrasting or harmonising colours that will enhance the message

- Be careful in basing a distinction solely on red vs. green; those in the audience who are colour-blind might miss it

- Choose colours so that the room can be illuminated as much as possible

Clip-art (prefabricated pictures) can be useful to make visuals attractive, assuming that the clip-art supports and emphasises the text message; otherwise, the clip-art might just distract. To produce a desired illustration, you can ungroup and make changes in a clip-art or combine different ones, but first check any restrictions on use that came with the clip-art.

Examples of visual displays (made in PowerPoint®) depicting the same message but composed with different layouts and backgrounds are shown in Figure 10:3. Figures 10:3a-c were produced using a presentation design template (provided with the software) for the background and an autolayout for the content, whereas Figure 10:3d was composed by writing the text on a blank page with no fixed layout, and by inserting clip-art.

Note that the displays in Figure 10:3 are in *horizontal format* (landscape orientation). We recommend this format because the projected image can be positioned high on the wall screen for better viewing. The format also allows longer lines of text, so that fewer lines are required to complete a point; this makes reading at a distance easier. Make sure, though, that your display is not wider than the screen when projected.

Use a font size that can be read from a distance

It is essential that everyone in the audience, including those sitting at the back of the room, can see and read your visuals. You should never need to say to your audience: "You probably can't see this, but …". If that is what you expect, don't use the visual! The font size might have been sufficient in a small room, or on a large screen, but always be prepared for the fact that the screen may not be large enough. Before completing your visuals, check the readability of a few of them under various conditions unless you

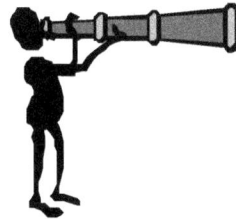

Use a large
font size

are certain of the font size needed. When using PowerPoint®, for example, you can also look at your slides in *slide sorter view* (select zoom 100%). If they are hard to read, your font may not be large enough.

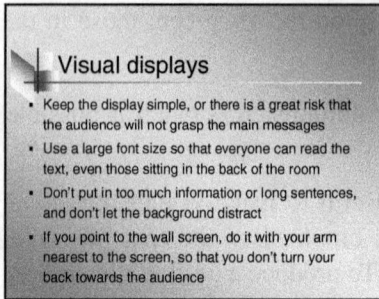

Figure 10:3a. Overloaded with text.
Poor contrast between text and
background. Light diagonal distracting.
Presentation template used.

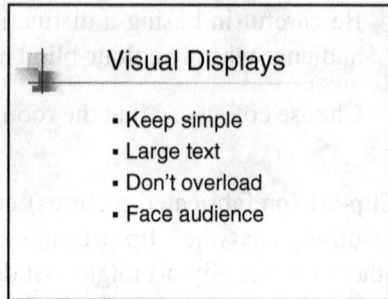

Figure 10:3b. Large brief text (to be
supported by the speaker's words).
Good contrast between text and
background. Autolayout used, but text
moved to centre

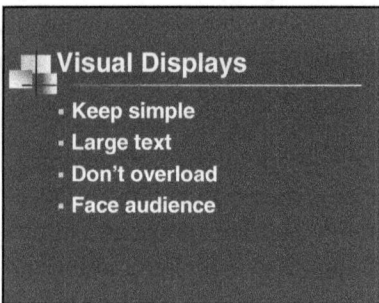

Figure 10:3c. Dark background, and
light and bold text for good contrast.
Autolayout used – text not moved to
centre.

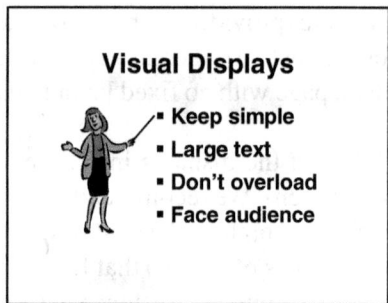

Figure 10:3d. Starting from blank page
– no fixed layout. Text bold. Good
contrast. Clip-art related to content.

Figure 10:3. Examples of visual displays depicting the same message.

The standard layouts offered in presentation software are intended to make it easy for you to balance the display and to use a proper font size. Note, however, that you might still need to change the line spacing or to move the text to get balance in the visuals. It might be easier, therefore, to use the option "blank page"; then you can take a more active role in designing your visuals.

When writing text on visual displays you should think of the following:

• Each display should be simple and easy to understand quickly. It is usually better to produce two simple displays rather than overloading one.

- Bullets can accentuate the text. Use symbols, such as dots, squares, circles, hands, or arrows; you can also use picture bullets or numbered bullets. Make a hard line-break (press Enter) to write text in a new bullet, but make a soft line-break (press Shift-Enter) to force text within a bullet to appear in a new line. Then at any time, you can change line spacing easily, both within and between bullets.

- A large font size must be used for the text, usually 24 points minimum; headings preferably larger. In standard layouts, the font size is about 30 points for text and about 45 points for headings.

- Bold text might be seen better, but only if the text is brief.

- Words written in lowercase letters (or with an initial Capital) are easier to read than words in all UPPERCASE.

- Fonts without serifs (e.g. Arial, Helvetica and Verdana) are often said to be easier to read in short text messages, whereas fonts with serifs (e.g. Times New Roman) are easier to read in full paragraphs. If you want a less formal look, use e.g. Comic Sans. For more advice on fonts, see the section "Electronic presentation".

Show a table or a figure?

Research results are often presented in tables in the written scientific paper. In visual displays for oral presentation, however, tables are often converted to figures, which usually makes it easier for the audience to grasp the message quickly. If the precise data are important, use a table. If the major purpose is to show a trend or to make a comparison, however, a figure can be more instructive. Below the same data are illustrated in a table and in a figure (Figure 10:4). Which will best help the audience to get the message?

Yield (kg) in different years

Year	Variety A	Variety B
1970	75	50
1980	275	150
1990	250	140
2000	375	175

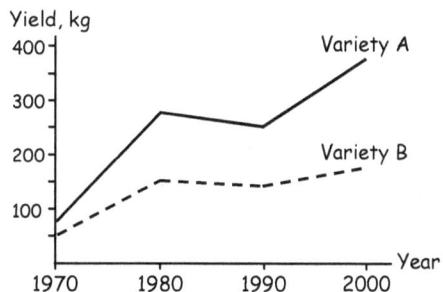

Figure 10:4. The same data illustrated in a table and in a figure.

As with text displays, both tables and figures must be simple and not overloaded. The time available for the audience to view each table or figure will be limited, so make sure that the message is clear. Abbreviations should be avoided, but if you have to use them, then make them logical or suggestive. Help the audience to understand the conclusion(s) for each table or figure. You can, for example, display a "take-home message" for each table or figure.

Tables should not be complex. Three or four rows and/or columns are usually enough. The font size must be as large as for text displays. Rounding numbers adds to clarity. The table should not be a copy of the full table in the paper, but could be a part of it.

Making your figures attractive should not be a problem, if you use presentation or spreadsheet software. You can use different types of charts depending on what you want to illustrate (see the chapter "Tables and Figures"). If possible, instead of having explanatory legends outside the graph, put a label next to each line or pie-segment. That will minimise having to look in more than one place to get the necessary information, and it will make it easy to see what a line or segment represents.

PERFORMING THE ORAL PRESENTATION

A presentation can be done in many ways,
as long as the audience interest stays,
but always be keen,
that what is said is heard,
and what is shown is seen.
Remember the communication ABC,
and enjoy the opportunity with an
audience to be!

An oral presentation can be performed in different ways. You can choose which type of visuals to use (assuming the necessary equipment is available), and also to what extent you use them. You can stand by the lectern, you can stand away from it to use more of your body language, or you can even move around a little. During a presentation it is usually better to stand, but there might be a few occasions when you could sit down, assuming the audience can still see you.

Choose an approach that fits your personality and that makes you feel comfortable. At all times, however, remember the people in the audience. You need to maintain eye contact with them throughout your presentation.

Keep eye contact with your audience!

An oral presentation should fulfil the communication ABC, i.e. be Audience-adapted, as well as Accurate, Brief and Clear. This, however, is not the whole story. Some additional basic requirements need to be fulfilled as well:

- Show interest and enthusiasm

- Keep eye contact with your audience

- Speak so that you are heard and understood

- Only use visuals that are clearly seen and that support your talk

Your own commitment is fundamental for a successful presentation. Showing interest and enthusiasm should not be difficult. You have a message of value to the audience! Focus on the message and on getting it across. Establish eye contact and make your listeners feel that you are talking directly to them; you will then also feel their response. Don't let it disturb you if someone in the front rows falls asleep; maybe he or she just had a late night! Concentrate on those who seem interested, but try to look in all directions. Don't concentrate so much on your visuals that you lose contact with your audience. Because eye contact is so essential, ask the organisers to keep the room light on.

When using visuals in a presentation, it is important to think of the following:

- If you want to point at something on the projection screen, then make sure you don't turn your back on the audience. It's important to face them, so hold the stick or laser pointer in your hand nearest the screen.

- If you give an electronic presentation, don't stare at the computer screen. Just have a quick glance at the screen when needed, and immediately re-establish eye contact with your audience. Practice switching slides, bullets etc., without interrupting the contact with the audience. You can stand away from the computer if you use a remote control or a cordless mouse, and you can glance alternately at the computer screen and the projection screen to keep track of your visuals.

- If you use an overhead projector, cover the surface while you are not using the projector for your presentation, to avoid the audience being distracted by a bright light on the screen, and because some people are sensitive to bright light. You might also cover the surface of the projector not covered by the transparency. Avoid using a sheet of paper, however, to release the content of the transparency in stages. Audiences often don't like it, and you get tied to the projector.

- Make sure your body doesn't block the audience's view of your visuals, especially if you point directly at a transparency on the overhead projector. It is better if you stand back close to the screen, but still make sure you don't block the view.

Speak to be heard and understood

Your voice is important in your presentation. It is essential that everyone in the audience can hear you. Think about your speed of delivery. If you talk too fast, then many people will not understand what you say, particularly if you don't speak in their native language. If you talk too slowly, however, then the audience's attention might decline. To maintain attention and interest, therefore, vary your voice level and speed, and now and then make short pauses. At all times, though, your speech must be clear, so speak up and articulate!

Speak clearly, but don't shout

If you use a microphone, then talk in conversational tone; the loudspeaker volume can be increased if needed. Practice using a microphone, if you are not used to using one. Avoid wearing a necklace or other hard items near the microphone; it might easily cause a disturbing noise. Dress so that you have a pocket for the battery, in case a wireless microphone is used.

Is a manuscript needed?

Whether you use a manuscript for the presentation depends on what makes you feel comfortable and on how much you have rehearsed. In many situations, it is useful to have some kind of manuscript.

You should *not* read your talk from a fully written manuscript; reading makes it hard to maintain eye contact with the audience. It is usually more difficult to understand what you are saying if you read from a manuscript than if you talk directly to the audience. If you must read from a manuscript, because of language difficulties, then learn to read whole phrases or sentences and face the audience while speaking, so that you maintain eye contact as much as possible. It may be easier to look up from a full text manuscript if you write with a large font size and put text only on the top half of each page.

A manuscript with *key words* to aid your memory is often a good solution. If your presentation is based on visuals, it can be helpful to have a reduced copy of each visual, together with key words to remember what to add. You might write the key words on each copy and place the words exactly where they are to be used, or you might use the notes function in your presentation software. With a copy showing the full slide, you know what will appear after the next click, if you use animation in an electronic presentation. If you plan to stand away from the lectern, then it can be handy to have your "script" on small sheets or cards. Be sure, however, that the key words are written in a size that you can read easily and that the pages are numbered, just in case you drop them during your talk.

Prepare and write carefully the introductory words of your talk. It is also helpful to have the closing words in written form.

Keep to time – rehearse

Rehearsal is often the key to a successful oral presentation, and probably the only way to make sure that you will stay within the time allotted. Considering the labour spent on planning the content and preparing visuals, it would be unfortunate not to spend the time needed on rehearsing. Rehearse on your own or ask a colleague to listen to you; do whatever feels comfortable. You can also use video or a tape recorder. If you are preparing your first major presentation, try to rehearse in front of a group of friends or colleagues. It is possible to rehearse without getting tied to an exact text – as long as you don't read from a written manuscript.

It is important to keep to the time allotted for your presentation. If you don't, the person chairing your session might terminate your presentation, otherwise the entire meeting programme will get "out of synch". Without proper time planning, therefore, you might have to stop your presentation before getting to the most essential parts.

The best way to keep control of time is to practice the presentation using a timer. Prepare a presentation that will last about 80 percent of the allotted time, because it usually takes longer to give the actual presentation and you don't want to feel rushed. Furthermore, prepare your presentation so that you can be flexible. It might happen that you actually have less time for your presentation than you expected, so decide in advance which parts to omit, if necessary.

Feeling nervous is natural

Feeling nervous just before and at the start of a presentation is natural; even experienced presenters have that feeling. Don't worry about it; adrenalin focuses your mind! Remember, you are the expert on your topic; nobody in the room knows more about what you will say than you. The best way to cope with nerves is to rehearse, to be well prepared, and to know what you are going to say (not word by word, of course). Memorise the first sentence of your talk; this will get you off to a good start, and thus ease the rest of your presentation. Once started, you usually don't suffer from nerves any more. Having good visuals, plus some key words for your memory, will keep you on the right course.

Go through your presentation once more before your delivery. Visualise that it works well. Inspect the room in advance, and decide where you will stand when you speak. Concentrate for a few moments before you walk up to give your presentation – and think about the privilege of having an audience interested in listening to your message!

Coping with questions

In most scientific meetings there will be an opportunity for the audience to ask questions or to raise issues for discussion, either after each presentation or in the general discussion part of the session. Some of these questions can be predicted, so you can prepare answers. You can even prepare some extra visuals with additional information.

Questions show interest, so let the audience understand that you appreciate their questions. Writing down a few key words from each question might help you to remember it, especially if there are multiple questions. It might be a good device to repeat each question briefly before you answer. That device lets the questioner know you understood the question, it ensures that everyone in the audience hears the question, and it gives you some time to think of an answer. Occasionally it can be difficult for the speaker to hear or understand a question; you can always ask the questioner to repeat or clarify it.

Your answers to the questions should be short; leave room for more questions. Don't give a new talk! If you don't know the answer, say so, and offer to find out. Avoid a dialogue between yourself and the questioner; direct your answer to the entire audience.

11

POSTER PRESENTATION

Poster presentations of research results are used increasingly at scientific meetings and on other occasions. Including a poster session as an alternative to oral presentations means that the number of papers/abstracts can be increased.

Research results can be presented effectively in a poster, and posters can have several advantages for both the presenter and the viewers:

- Main messages can be highlighted

- Viewers can study the information at their own pace

- There is an opportunity for questions and meaningful dialogue between presenter and viewer

- The poster can be reused, e.g. at the presenter's home institution

An effective poster takes time to prepare. Meeting organisers, therefore, have a responsibility to plan a poster session that allows both presenters and viewers to get the most out of it. It is important that posters are located where viewers can find them easily, and that sufficient time is allotted in the programme for poster presentations, which should not be at the same time as oral presentations in the same scientific area. A poster session without an opportunity for meaningful dialogue with the author(s) precludes a major benefit of this form of scientific communication.

ATTRACT VIEWERS AND SHOW THE ESSENTIALS

Poster sessions are often filled with a large number of posters, and there is intense competition for viewers' attention. You need to attract viewers and make them curious or interested enough to look at your poster, but be aware that you have only a few seconds to achieve this. The wording of the title and the overall appearance of the poster, therefore, are of utmost importance. To maintain and further arouse interest, your poster needs to have a brief, clear message, so that viewers can quickly understand the most important points. Later, viewers can look for more details and possibly discuss the topic with you.

The poster should tell why the topic is important, the objectives of the study, what methods you used, the most important results, the main conclusions, and the possible implications. The methods normally are mentioned briefly, whereas the results (with emphasis on visuals) and the conclusions form the largest part of the poster. If space is short you may omit the introduction, but do not omit the objectives. You might want to include a lot of information, but remember that viewers will miss your main "take-home" message if you overload the poster. Extra information can be given in a handout.

As with an oral presentation, it is important to adapt the poster to the audience. Anticipate their probable questions, both to decide the content of the poster and to prepare answers to questions during the poster discussion.

DESIGNING THE POSTER

Before you start to plan your poster, you must know its maximum dimensions for height and width, as specified by the organisers. Make sure you get it right, so that you don't design a poster in landscape orientation when it should have been in portrait. The organisers may also set rules on the structure of the contents, but usually you may do it in the way that you find best. It is important to have some consistency within your poster, but variation between posters can add

Check poster dimensions before designing the poster!

substantially to the interest of a poster session. So use your imagination to make an attractive and informative poster, and remember: the key to an effective poster presentation is simplicity.

Choose layout and content

The poster content is often laid out in sections under headings such as Objectives, Methods, Results and Conclusions, but you might also use more informal headings, such as short statements or questions. Novelty might help make the poster more attractive.

Before deciding on the poster layout, it might be helpful to set up a one-page model in proportional scale, preferably on a computer. The content can be arranged in columns or in rows, but you can also choose some other structure for the layout, e.g. circular or counter clockwise. If the poster is wide, then it might be best to arrange the information in columns, so that the viewer walks along the poster from left to right, especially if many people can be expected to view the poster at the same time. Whatever structure you choose, place the title at the top of the poster, followed by the author names and addresses, and use the same title and number as in the meeting programme.

Your poster should be self-explanatory. Arrange the content in a logical order, and remember that the most important messages on the poster should be placed where you think the audience will notice them best. Don't hide the conclusions at the bottom of your poster; rather place them at eye-level for the audience. The conclusions can also be highlighted.

Visuals, such as tables, figures, photos, and other illustrations (e.g. drawings and clip-art) can make the poster attractive and easy to understand, but be sure they help tell your story. Strive to find a balance between text and visuals, with regard to both size and proportions of the poster. Some examples of poster layouts are given in Figure 11:1.

A poster needs a unifying background, which separates it from the poster board and neighbouring posters. To achieve this, you can mount individual elements of the poster on coloured cardboard ("multi-part poster") or, alternatively, you can produce a "single-sheet poster". These alternatives are dealt with in the next section. Remember that the background must not be distracting; it is the message that should be emphasised. If you want to unify groups of data on the poster, then you might use a sub-background colour that harmonises with the main background. If you want to highlight part of the poster, then you might choose a contrasting colour. Text is usually read best when written on a background that is light (e.g. light beige or grey), but not pure white.

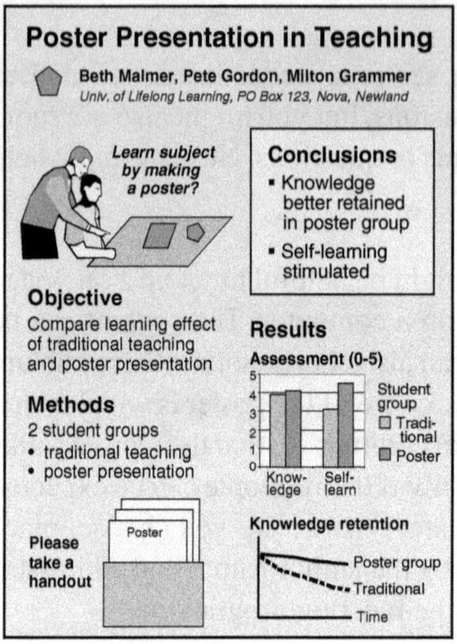

Unifying background. Conclusions at eye
level. Illustration to draw attention.
Handout provided.

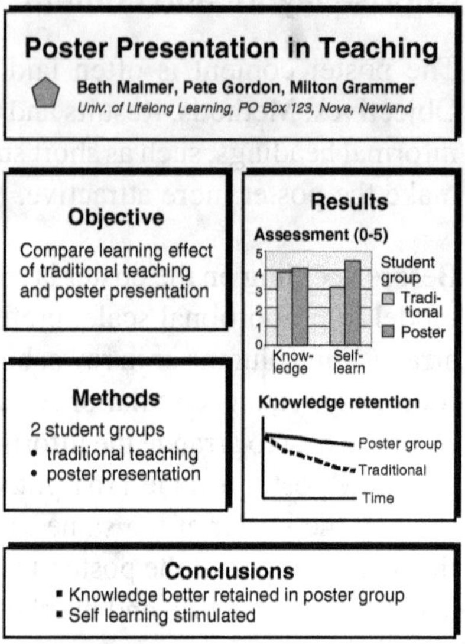

Distinct sections, but no unifying background. Conclusions far below eye level.

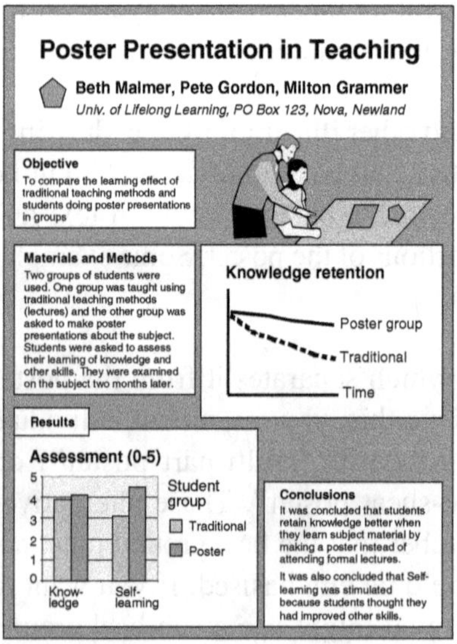

Unifying background. Text too small.
Visuals larger than needed. Results not
together. Conclusions "hidden".

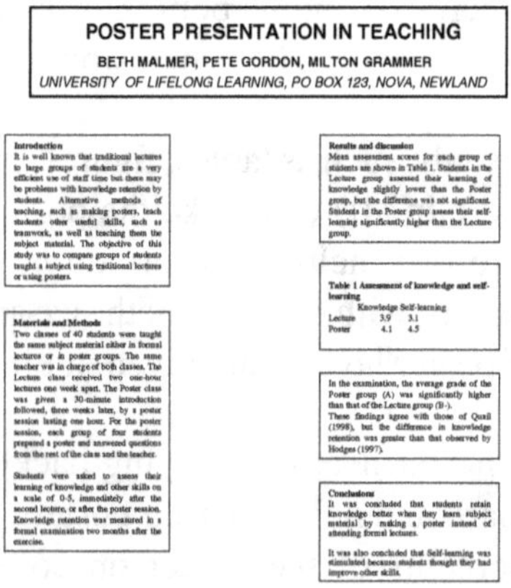

Enlarged manuscript mounted on the
poster board. Should not be accepted!

Figure 11:1. Examples of poster layouts.

Occasionally you see posters that are composed of a number of individual sheets mounted directly on the poster board, often with a dark frame around each and empty space between the sheets. This composition usually gives a disjointed impression, which can be avoided with a unifying background sheet.

Each section of the poster should contain just a few important messages. You do not need to write complete sentences, so delete most of the words, but leave the meaning. The details can be given in a written paper. For example:

Description of Materials and Methods in printed summary for handout:

Two hundred people, who regularly attended the Heart Clinic at the Royal Free Hospital, were randomly assigned to two groups of 100. One group was asked to consume two eggs per day for thirty days, and the other group acted as a control. A blood sample was collected from each patient at Days 0 and 30 for analysis of plasma cholesterol (Smith, 1996). Patients were weighed at the time of blood sampling.

Description of Materials and Methods in the poster:

Treatments (30 days duration)
A   2 eggs per day (n=100)
B   Control (n=100)

Measurements (days 0 and 30)
   Plasma cholesterol
   Body weight

Don't overload the poster. Leave some empty space, so the viewer can understand the content easily; but make sure that the text and figures don't appear unrelated. Remember the purpose – to awaken interest and stimulate discussion.

Make the poster

You can create your poster on a computer, either partly or fully. Before creating the individual elements of the poster, however, you should decide how it will be mounted. A poster can be produced in several ways, two of them are:

- *Multi-part poster*

  A "multi-part poster" is where the individual elements are produced separately (usually on a computer) and mounted on a unifying background paper or card. To make the poster stable, the background paper can be pasted onto cardboard or foam-core board. The poster can be given "life" and a deeper dimension if some parts, e.g. illustrations and graphs, or some demonstration materials are attached so that they stick out a little from the background.

A multi-part poster often needs to be split into segments for easy transport. The final mounting is then done at the meeting site, where the segments are joined with wide tape on the back. The poster can be completed in large parts while still at home. Save some text sheets or illustrations to mount at the meeting, however, so that you can hide parts of the joints. The final mounting is easier if the segments are taped and folded in pairs before transport.

- *Single-sheet poster*

  If you want your poster to be produced as a single sheet, you can create it on a computer, and then print it on a special printer to achieve the desired size. The single-sheet poster usually is printed on paper and might also be covered with plastic laminate. To transport the poster safely, you might need a poster cylinder.

It is not easy to say which type of poster will be better. The multi-part poster might be more "alive" than a single-sheet poster, but be prepared to spend time at the meeting site on the final mounting. The single-sheet poster, however, is simple to mount at the meeting site, but the printer needed to produce it might not always be available, or it might be expensive to use. The ultimate preference is largely a matter of taste. What you should never do, however, is just to enlarge your written paper to form a poster; this is guaranteed to look unattractive and unprofessional, and people will not waste time reading it.

When making the poster, it is also important to think of some visual aspects for formatting:

- Colours will enhance the poster, but too many colours will distract or give an uncoordinated effect. The title and headings can be attractive in colour, but the body text is usually easiest to read in black (or dark blue). Colour can be used to highlight, separate, or associate information. Think of the background colour when you choose colours for headings. Remember that colours on the computer screen might not look exactly the same in print.

- Bullet points are easier to read and to understand than long paragraphs of text. You can use standard bullets, or insert symbols or pictures as bullets.

- The font should be easy to read and can be either with serifs (e.g. Times New Roman) or without serifs (e.g. Arial and **Comic Sans**). Use a font with proportional spacing between characters, rather than one with fixed spacing (e.g. don't use Courier). Bold letters in the title and headings can facilitate reading from a distance. Words in lowercase letters (or with an initial Capital) are easier to read than words in all UPPERCASE letters.

- Text size must be large. If the room is crowded it might be difficult for the audience to come close to the poster. The poster title should be large enough to read from a distance of 3–5 m and the text from 1.5–2 m. The font size of the printed poster, therefore, should correspond to about 110–120 for the title, 60–70 for headings, and 30–40 for the body text.

- Tables and figures must be easy to read and to understand (for examples of different types of graphs, see Figure 3:1 in the chapter "Tables and Figures"). Use an appropriate font size, limit the amount of information, and help the viewer to understand quickly what the table or figure is about. A written take-home message near the table or figure might also help.

- Clip-art can be useful to illustrate the poster; modify or compose clip-art if necessary to fit your purpose. You might also put an "attention-getter" above or below the poster. The attention-getter could be made in cardboard, for example, or it could be a striking photograph. A photo of the presenter can be placed near the poster title to help the audience to identify the author for questions and discussion.

- A matt poster surface is usually preferable to a glossy one, because light reflecting from a glossy surface can make your poster difficult to read. For the same reason, photographs should have a matt surface.

PRESENTING THE POSTER

Don't forget to bring what you might need for the final mounting of the poster, e.g. push pins, glue, spray-adhesive, tape, adhesive dough. You might attach a small holder containing your business cards, for those who want to contact you later. It might also be a good idea to produce a one-page handout and place copies near your poster, so that those who are interested can take one. Such a handout could include on one side a reduced copy of the poster, and on the other side the names of authors, (e-mail) addresses, the abstract and relevant literature.

Make sure you arrive on time for the poster session, and stay by your poster for the entire session. You might want to prepare a 2–5 minute presentation to quickly guide interested viewers through your poster. Remember, though, that for the most part the poster should be self-explanatory. Your major role is to be prepared to discuss your topic, to respond to questions, and to provide additional information to viewers if needed. A folder with additional information, e.g. tables and figures, can be useful for this purpose. Discussions during a poster session are different from those during an oral presentation session. The poster discussion is more detailed and more on a one-to-one basis; it's a dialogue that also gives you a splendid opportunity to establish valuable contacts!

12

TRAINING STUDENTS IN WRITING AND PRESENTATION

Children learn to communicate from birth: normally they learn to speak within three years and are taught to write in school. The development of communication continues through adolescence, so that when students enter university they have plenty of experience in oral and written communication. Emphasis in pre-university education is on telling stories and writing expressive prose or poetry. At university, however, more marks are awarded usually for the number of facts recalled than for writing quality. Style is the main thing that distinguishes scientific writing from writing fiction, because science demands precision and accuracy. To demonstrate their true potential, students not only need instruction in their chosen subject, but they also need training in the techniques of scientific communication.

Communication skills are transferable, so if these skills are integrated into a degree programme at an early stage, then students will benefit throughout their studies. Communication skills are also beneficial in most careers and are often a key to success when students seek employment after graduation. In a survey of skills required in graduates, both employers and graduates listed communication as most important.[1]

Training and practice
- a key to communication skills

Students can be trained in writing and presentation by using examples that are relevant to a specific subject. Students can thus learn communication within most subject courses, which will enhance their overall learning experience. There is a Chinese proverb that goes: "I hear, I forget; I see, I

---

[1] CLUES (1998) Personal communication: CTI Centre for Computer Based Learning in Land Use and Environmental Studies, Aberdeen – Skills For Life Project.

remember; I do, I understand". Every teacher knows that the best way to learn a subject thoroughly is to explain it to someone else; similarly, the best way for the student to learn the subject is to "do" it. Integration of communication skills into the curriculum, while using topics that are relevant to the subjects studied, therefore, should be seen as an effective way for students to learn material in depth.

UNDERGRADUATE AND MASTERS LEVEL

Communication requirements of undergraduate students change as students progress through their degree programme. These changes often coincide with a decrease in class size as specialisation increases. Initially, students have to communicate ideas and facts through essays and reports. Oral communication might be limited to informal occasions, such as asking questions in lectures and discussing practical classes with an assistant. Later, students can be trained to develop presentation skills when smaller class sizes make this more practicable and less daunting.

General study skills

All undergraduates need to learn basic study skills, such as managing their time, using libraries and computers, writing essays and practical reports, working in groups, and preparing and presenting seminars. These skills should be introduced at an early stage of a degree programme, and can form part of the induction or orientation process. Students can be given a pack that contains advice on important study skills, covering material such as:

- Managing time

- Learning in tutorials (teacher plus 1–6 students)

- Making notes

- Finding information (library skills, literature search, Internet, etc.)

- Reading efficiently

- Writing reports

- Writing essays

- Working in groups

- Giving effective presentations

- Studying for exams

- Answering exam questions

It is even more helpful if a member of staff goes through the information with groups of 10–15 students. Teachers can emphasise that some of the material covered is not relevant immediately, but it is included at this stage because the time when individual skills are required varies among subject areas and because it is better to have all material on study and communication skills gathered in one document. This type of induction course requires intensive staff time, but it is considered to be a valuable use of resources because it saves individual remedial work later.

Training communication skills throughout the degree programme

Integration is the key to successful training in communication skills, and students learn most effectively by practising these skills throughout the undergraduate degree programme. It is vital to give students feedback on their performance, so that they know what they did well and how they might improve (see the chapter "Reviewing Papers and Presentations"). In addition, essays, reports and other forms of written coursework should be marked to an agreed convention, such as that shown in Table 12:1. In the early stages of a degree programme, training in communication skills and feed back might be on work done in groups, whereas in the later stages the training can be more focused on individual students. Students can be given broad training in presentation of scientific ideas through oral and written communication and through poster presentations, although the approach might differ among subject areas. This alternative to traditional lectures makes learning more enjoyable and effective for students, and can make life more interesting for the lecturer.

Seminars

A seminar delivered by a student or a group of students is probably the most common alternative to lectures. Students might work in a group of four or five, for example, and prepare an oral presentation on a topic relevant to the theme of the module. Preparation time might be about six weeks, and the group presents its talk to the class with the teacher present.

Table 12:1. Example of a marking convention for written work submitted by undergraduate and masters students

Performance	Marks range (%)	Comments
Outstanding	90–100	Outstanding, excellent or very good structure and arguments. Evidence of
Excellent	80–89	critical thinking and understanding. Comprehensive and correct
Very good	70–79	knowledge. Use of relevant examples. Evidence of wider study well beyond set classes.
Good	60–69	Good structure, some critical thought, some errors and omissions. Usually few conceptual errors, and some evidence of wider study.
Satisfactory	50–59	More errors or omissions, adequate structure, but limited argument. Factual information and detail limited; no evidence of wider study. An understanding of the question should be evident.
Barely satisfactory	40–49	Poorly structured with little evidence of critical thought or argument. Significant factual error and no evidence of wider study. A basic answer indicating a limited understanding of the question.
Weak	30–39	Some indication that the question has been understood, but much lack of structure and information. Extreme omission and factual error.
Extremely weak	20–29	Limited evidence that the question was understood. Extreme superficiality.
Inadequate or irrelevant	0–19	A response so flawed or superficial that a complete lack of understanding is evident.

Input from the teacher depends on the quality of students. All students usually need some guidance at the planning and preparation stage. Better students require little guidance or involvement during the presentation. Poorer students, however, require guidance and assistance, either by the teacher acting as an informed member of the audience (asking directed questions) or by the teacher correcting factual inaccuracies or omissions.

As with written work, it is vital that the teacher gives constructive feedback on the performance, whatever the standard of the oral presentation. Feedback should be positive and encouraging, particularly when given publicly. Giving your first presentation is always traumatic. The teacher will usually feel that he or she could have done it better than the students, but the experience of making the presentation is as important as learning the subject content, and students should not be discouraged by adverse criticism. It is

Give constructive feedback to students

also important that students have been properly taught on what is essential in an oral presentation (see the chapter "Oral Presentation and Visual Displays").

Oral presentation of scientific papers

An extension of the seminar can be a journal club, where students concentrate on a single paper published in a scientific journal. This activity is best performed in groups of three or four students. Each group selects from a refereed journal one paper that was published in the past three years. The first task is to write an abstract of the paper in the students' own words, according to rules for writing a scientific paper, and submit this for assessment. The second task is to prepare and give an oral presentation of the paper under conference conditions, e.g. ten minutes for presentation plus five minutes for questions. A teacher should chair the presentation session and ensure that papers run within the time limits. To ensure adequate questions from the audience, each group of students could shadow one other group by reading the paper that group will present. Students enjoy this exercise and it helps them to learn about the structure of scientific papers, to write abstracts, to stay on time, and to deal with questions, as well as to acquire oral presentation skills.

Writing and presenting a review paper

The seminar and journal club can be combined so that students write and present a review paper on a specific topic. The task can be done individually or in groups, with some supervision from a teacher. Students might summarise several scientific papers, or they might perform their own literature search and write a comprehensive review paper. When a review paper is presented orally, two or three students can be given the task to critically assess the paper and the presentation in a constructive manner. Thereafter, a teacher can add further points of assessment (see also the chapter "Reviewing Papers and Presentations").

Writing popular science

As discussed in the chapter "Other Types of Scientific Writing", popular science articles require a less formal structure than scientific papers written for journals. The ability to interpret scientific papers for a non-specialised audience is a valuable skill for graduates later in their careers.

Students can be trained to summarise scientific papers and to rewrite them in a style suitable for the popular press. To make the exercise more valuable, a journalist could be asked to give feedback on the articles, and some articles might even be published in magazines.

Posters

An important method for communication of scientific ideas is the presentation of posters at conferences and in industry. Presenting a poster can be part of training in communication skills. Students in groups of three or four research a topic relevant to their course, and produce an abstract and a poster to illustrate their findings. These posters are displayed in a session where staff and students view the posters and ask questions of the presenters. Before this exercise, students should be taught about poster design (see the chapter "Poster Presentation").

Undergraduate and graduate theses or dissertations

In some universities, undergraduate students conduct a practical research project over a relatively long period of time; masters programmes almost always include a research project. These projects might be linked to an on-going research project but, ideally, students should have some flexibility in their choice of research area, and they should help to design the experiment. The research project normally contributes a substantial proportion of the marks for the final degree. Under the guidance of a supervisor, each student performs a literature review, conducts an experiment, interprets the data and writes a dissertation. In addition, students often present results of their research to fellow students and staff, either orally or as a poster. A more formal presentation might have to be made in the presence of an external examiner. A formal presentation differs from an informal group presentation, because in this formal presentation students work individually to present a topic with which they have been intimately involved.

POSTGRADUATE LEVEL

The training of postgraduate (PhD) students in communication skills normally is done on a more individual basis than of undergraduates and masters students, but there is some scope for teaching basic skills to a group of students. Postgraduate students normally will get more detailed training in scientific communication than that provided undergraduate and masters students. Two of the most vital skills to encourage in postgraduate students are

Postgraduates need planning skills to use time efficiently

planning and organisation. Without planning and organisation, students will not be able to complete their studies on time, and will have nothing of value to communicate.

PhD courses on scientific writing and presentation

All postgraduate students should attend an in-depth course on writing and presentation skills, and be required to read some literature on the topic. The course should include instruction on scientific writing, advanced literature searching, managing references, organisation and planning. The course should also include discussion of what makes a good presentation, as well as opportunities to engage in practical sessions, which can be video recorded so that students can observe their own performance.

In a course on scientific writing, students could work in groups to review critically the content and style of some published papers. Groups can be paired and given the same paper to reivew; this approach provides "friendly" competition between groups within pairs. Groups of students could also be given a full paper and asked to write an abstract that is different from that in the original paper. Alternatively, individual students within a group could write different sections of a paper, which could be combined and discussed by the whole group. Whatever the approach, practical sessions work best with groups of students sharing ideas.

In a course on oral presentation, every student could be asked to prepare and present a five-minute oral presentation on an optional or specified topic. The presentations should be video-recorded, if possible, including presenters answering one or two questions. It might be good to let each member of the group deliver his or her presentation, and then view the videotape one presenter at a time. The presenter should comment on the performance before other students and the teacher give their comments. It is important that comments are constructive and that positive aspects are highlighted. Students learn a lot from this exercise, both by watching and discussing their own performance, as well as by watching and discussing the performance of other students.

Presentation skills

Informal presentations

Postgraduate students should be encouraged to give informal presentations of their research work regularly to staff, visitors and fellow research students. The emphasis of these presentations is not so much on the actual presentation, but more on the discussion of the work. In this way, students start to feel ownership of their work and get used to thinking about ways to explain their results to other people.

Formal presentations to the department

Formal presentations are sometimes required as part of postgraduate assessment schemes. During their course of study, postgraduates may present their work to the whole department, either orally or by posters. The first presentation might be six months after registration, and consist of a literature review and outline of the project. The second presentation might be about twelve months later, when students should have preliminary results to present and discuss. The third presentation might be when the project is nearing completion, so that students can be expected to give a professional presentation of the whole thesis. Ideally, the audience should include staff and students from different disciplines, so it is important that the presentation be simple and clear.

Formal presentations at scientific meetings

All postgraduate students should expect to make a presentation at a scientific meeting. Because students are representing their university in public, it is essential that they be well trained and perform professionally.

For oral presentations to be given at scientific meetings, preparation should start about two months in advance, when students present an outline to colleagues. The main purpose of this outline is to confirm the content of the presentation and assess the visual aids, which may be in draft form. The outline will probably be modified and repeated after about three weeks. On this occasion, timing of the presentation is studied, and content and visual aids are finalised. Once the revisions have been made, there should be at least two more rehearsals to fine tune the presentation and to think of likely questions that will be asked. If this sort of rehearsal programme is followed, then the conference presentation will be much better.

For poster presentations, drafts should be produced and assessed by colleagues. The final poster can be displayed, and students should be questioned on the content.

Writing skills

Apart from the initial training course that should be offered to all new postgraduate students, training in writing skills is almost entirely on an individual basis, between student and supervisor. Training usually parallels the development of a student's scientific ability. In the early stages of a PhD programme, considerable help with style is offered. In the later stages, however, the student requires less help with style, leaving more time for discussion of content and for development of ideas. The role of the supervisor changes from that of a mentor to that of a colleague.

Writing progress reports

An annual progress report, as well as a formal presentation, might be required as part of a postgraduate assessment scheme. This report does not have the formal style of a scientific paper, but students can be expected to produce a well-structured document that adequately summarises their progress. Students normally discuss drafts of the report with their supervisor before submission.

Research sponsors, including research councils, government agencies and companies, usually require progress reports on an annual basis. Postgraduate students normally are expected to have a high degree of input into these reports, especially in the later stages of their PhD programme.

Writing the PhD thesis or dissertation

The ultimate goal of a PhD student is to write a thesis or dissertation that is acceptable to the examiners for the award of the degree. In many countries, PhD examinations are based upon a written thesis or monograph of 150–300 pages. In other countries, however, a series of 3 to 5 journal papers, plus a general introduction and a general discussion, are the usual form for a thesis. Students are expected to complete most of their writing before the end of their supervised study. This requires discipline in the planning stages of the project and in sticking to the plan. If various progress reports have been written diligently by the student and corrected by the supervisor, then much of the writing can be editorial in nature.

There is often a tendency for students to keep modifying the thesis to the point of trying to make it perfect, thus delaying it . Such perfectionism must be resisted because it will not be appropriate in later employment. The supervisor should also avoid overcorrecting the thesis in style and language, which also introduces delays; besides, the thesis is supposed to be largely the student's own work. If adequate training has been given throughout the PhD programme, then students should have developed sufficient skill in writing to produce a thesis with an acceptable style, so that the main focus of the supervisor's comments can be on content.

Although the goal of the student is to produce a thesis, the goal of the institution is to produce scientific papers that can be published in refereed journals. This is the main criterion by which the research quality of most universities is assessed. Scientific papers are also important for employment and career progression not only for academic staff, but also, increasingly, for industry staff. If the thesis is a monograph, then the student should draft one or two papers before graduation, for a number of reasons – it is good training for the student, the student should be the most expert person on the research programme, it saves the supervisor time, and the student might not have time to write papers after starting a new job.

TRAINING COMMUNICATION SKILLS - IN SUMMARY

- It is essential that undergraduate and postgraduate students be trained in writing and presentation techniques as part of their scientific education.

- Requirements for training in scientific communication change as students progress through their degree programmes, and methods move from training in general communication skills towards training in specific skills.

- A variety of approaches can be used to train students in communication. This variety not only gives the student a broader education, but also helps the student to maintain enthusiasm.

- Writing and presentation skills can be taught effectively using topics of relevance to subject-specific science courses. Such an approach will enhance the overall learning experience, without increasing contact-time of staff.

- Programme planning and writing practice are essential if postgraduates are to complete their thesis on time and be able to write papers for publication in a timely manner.

13

REVIEWING PAPERS AND PRESENTATIONS

The process of reviewing other people's written papers is often called "refereeing", to distinguish it from the process of writing a literature review, but you are doing the same thing – critically appraising the positive and negative aspects of someone else's work.

Whether you are reviewing written material, oral presentations, or posters, you should ask the same basic questions:

- Has the author conveyed his or her message clearly?

- Is the aim, hypothesis, or problem clear?

- Are facts presented clearly?

- Are references given where needed?

- Do the facts support the conclusions?

- Are the structure and layout correct and appropriate?

- Is anything missing or superfluous?

When reviewing the work of others, be aware of the effects that your comments may have on their self-confidence, particularly for students and young scientists and especially if it is their first attempt at scientific writing or public speaking. Always give feedback in a constructive manner; praise the good parts and try to explain alternatives to the bad parts, rather than just saying they are wrong. This positive approach is also more professional, especially when dealing with the work of more experienced scientists; you do not impress anyone by being totally negative about a paper or presentation.

WRITTEN PAPERS

When refereeing or reviewing a written paper for a journal, you need to consider not only how the topic is presented, but also whether the science is good. In addition to the basic questions listed above, you might ask yourself the following questions:

- Does the title of the paper accurately reflect the content?

- Is the abstract complete and does it stand alone?

- Is the subject material relevant to the journal?

- Is the experimental design appropriate for the hypothesis being tested?

- Are the methods appropriate and fully described or referenced?

- Is the statistical analysis correct?

- Are the results presented properly?

- Is material duplicated in text, tables, or figures?

- Does the author interpret the results correctly?

- Does the author describe non-significant differences as though they were significant?

- Could alternative interpretations or conclusions be drawn from these results?

- Does this paper make a significant contribution to the literature?

The journal might provide a referee with a checklist that contains additional points to consider. If the paper contains computations, then check some of them against the data provided. If the results do not agree, then ask the author to check all the data and results. Check that the written text is logical, coherent, and easy to read. Check also that each reference in the paper is cited in the reference list, and that each reference in the list is cited in the paper. This should be the author's job, but it is surprising how many papers are incomplete.

You will be asked to write a brief report about the paper and to give a clear recommendation if the paper should be accepted, returned to the authors for revision, or rejected. Make sure your report is constructive and gives sufficient detail about each criticism, so that the author can understand clearly what you require. Start with positive remarks, which will make it easier for the author to accept the critical remarks. Outline any general criticisms of the paper as a whole, giving references, if necessary, to support your arguments. The paper should have page and line numbers, so refer to the relevant part of the paper for each point. Move on to minor queries or mistakes, listing these in the order they appear in the paper. Finally, round off your report with a general conclusion about the value of the paper. If, for example, you state that the statistical analysis is incorrect or incomplete, then state in the conclusion if you consider it worthwhile reanalysing the data, or if there are fundamental flaws in the design. Remember that you have to uphold the standards of the journal.

Send your report promptly to the editor. Make your decision clear, not only in the report, but also in a covering letter (or e-mail, if submitted electronically). Be prepared to look at the manuscript again after the author has attended to corrections. Your arguments may be challenged (e.g. see the section "Editor's and Referees' Reports" in the chapter "Getting a Paper into Print"). Admit when you have made a mistake, but point out that the relevant section could be written more clearly if you, as a reviewer, found it ambiguous.

STUDENTS' WRITTEN WORK

The principles that apply to journal papers also apply to reviewing and assessing students' written work . With a journal paper your remarks might concentrate mostly on content, whereas with students' work your remarks might be about style as much as content. Remember that your feedback should form part of the student's learning process. Rather than indicating the faults in a paragraph, for example, ask the student what is the point being made in that paragraph, and where in the paragraph is that point made. Let the student do most of the talking; students learn most when they have to explain their written work.

Remember to give constructive criticism *and* praise! Focus mainly on general remarks in oral feedback, where you also can give examples of mistakes in style and formalities, but write detailed notes on the manuscript that you return to the student. You should also give some concluding remarks. If the written work is graded, always give your marks to an agreed standard (see the chapter "Training Students in Writing and Presentation"). Remember that

Give praise

marks on their own tell students nothing about *where* they went wrong; so always let them know *how* a text or presentation can be improved.

If you are a student reviewing another student's work, then follow the advice given above about feedback. In addition, you might consider what information might help you if *your* work were being reviewed, and make sure that is included.

BOOK REVIEWS

Book reviews call for a different approach than that used to review scientific papers. Not only do you need to comment on the content and style of the book, but you also need to assess the value of the book for its intended audience. Remember that the purpose of a book review is to help readers decide whether to recommend the book or even to purchase it. They need to know what is in the book; if the information is reliable, appropriate and complete; if the book is easy to read; and if the book will be useful to them. Read the book thoroughly and try to answer those questions in your review.

A book review should start with a general statement of what the book is about, who it is aimed at, and what you thought of it. You can then discuss in detail specific topics or sections of the book, giving examples where appropriate. Finally, finish with a conclusion about the value of the book.

You are asked usually to give your opinion on the book; this should be in relation to the intended audience. An introductory text on chemistry might seem too basic if you are a professor, for example, but could be at the right level for first-year students. As with any review, emphasise positive aspects of the book, as well as including criticisms. Make sure that criticisms are justified by evidence taken directly from the book; remember that others can read the book and check your assertions, so make sure quotes are accurate and in context. Do not focus on minor errors of grammar or style; if there are many minor errors, this can be indicated in a brief statement. Errors of omission need particular care. Before stating that something is missing, read the book or section again to ensure that you have not simply overlooked that material. Then decide whether the omission is important enough to be highlighted – does the omission significantly alter the message?

Book reviews are normally written in the first person, because you are stating your personal views. The prose can usually be more informal than that used in a scientific paper (see the section "Popular Science Article" in the chapter "Other Types of Scientific Writing"). By all means, introduce some light-hearted comments, but never try to make jokes at the expense of the author(s) or of the book; this usually makes you look silly. Some reviewers try to make themselves look superior to the author(s) by belittling the book or being unduly negative; this serves no purpose, except to show that the reviewer is unprofessional. Remember that you have a duty to the author(s), the publisher and the potential readers of the book to give an honest, fair, and balanced review.

ASSESSMENT OF PRESENTATIONS

Exercises in communication through oral and poster presentations can form part of the students' overall assessment, as well as provide training for them.

Oral presentations can be assessed using appropriate criteria:

- Structure and content of the presentation
- Clarity of delivery (speaker heard well, easy to understand)
- Enthusiasm and contact with the audience
- Design and content of the visual aids
- Use of visual aids
- Evidence of preparation (not reading a script)
- Keeping time
- Handling questions
- Quality and content of handout (e.g. abstract)

It is important that students and staff both are aware of the assessment criteria used. Students often perceive the delivery, or the talk, to be the main part of an oral presentation. It is acknowledged by assessors that nervousness can ruin a delivery, even by able and experienced students. Everyone suffers from nerves when performing in public, and the ability to overcome this is attained only with practice and encouragement. Fears can be allayed if the importance of *all* criteria is stressed; students should be awarded a substantial proportion of their marks for the quality of their visual aids and their handout. Like assessment of written work, it is important when assessing oral presentations to give constructive criticism *and* praise! Students should get help to build self-confidence and to learn how to improve their presentations. Students can also get help by reading the chapter "Oral Presentation and Visual Displays" before starting to prepare their presentation.

Problems of assessment might arise when students work in groups. Teamwork itself is a useful skill to acquire and develop, and individual presentation and assessment are often impractical with larger class sizes. In all teams, there are leaders and there are followers; sometimes groups of students are cross if one member of the team does little

Develop teamwork skills

work towards the presentation. One way around this problem is to award the team a mark and to ask them to negotiate individual marks according to their effort.

An experienced scientist might not always be the appropriate person to assess a presentation, at least not as the only person. Experienced scientists usually have preconceived ideas about the content, greater knowledge of the subject material and, with the benefit of experience, may have presented it differently. An alternative is to use a scheme of peer assessment, whereby each student in the audience marks the presentations according to the same set of criteria. Criteria should be provided on a sheet for each student marker, as a set of tick boxes or as a score sheet. This system has further advantages: it makes the audience more aware of the elements of a good presentation, and allows the audience to give an indication of how much they learned during the presentation. It is important that students are made aware of how the assessment criteria correspond to different grades, before they are asked to evaluate another student's presentation. In addition, a teacher should moderate the marks to ensure they comply with the university's standards. These sheets can be collated by the teacher and used to generate grades.

Poster presentations can be assessed in a similar way using criteria such as:

• Informative without being overloaded
• Attractive and easy to read from a distance
• Emphasis on conclusions
• Balance between text and illustrations
• Background and use of colours
• Presenter's handling of questions
• Quality and content of handout (e.g. abstract)

For details on what is important in a poster, see the chapter "Poster Presentation".

146

FURTHER READING

The following list of books and web sites might be useful when writing and presenting a scientific paper. The list, however, is not exhaustive.

BOOKS

Alley, M. 1998. The Craft of Scientific Writing. 3rd ed. Springer, New York.

Alley, M. 2003. The Craft of Scientific Presentations: Critical Steps to Succeed and Critical Errors to Avoid. Springer, New York.

Anholt, R.R.H. 1994. Dazzle 'em with Style - The Art of Oral Scientific Presentation. W.H. Freeman and Company, New York.

Booth, V. 1993. Communicating in Science: Writing a Scientific Paper and Speaking at Scientific Meetings. 2nd ed. Cambridge University Press, Cambridge.

Briscoe, M.H. 1996. Preparing Scientific Illustrations: A Guide to Better Posters, Presentations and Publications. 2nd ed. Springer, New York.

Davis, M. 1997. Scientific Papers and Presentations. Academic Press, San Diego.

Day, R.A. 1995. Scientific English: a Guide for Scientists and Other Professionals. 2nd ed. Greenwood Press, Westport.

Day, R.A. 1998. How to Write and Publish a Scientific Paper. 5th ed. Cambridge University Press, Cambridge.

Gustavii, B. 2003. How to Write and Illustrate a Scientific Paper. Cambridge University Press, Cambridge.

Huth, E.J. 1995. Scientific Style and Format: The CBE Manual for Authors, Editors, and Publishers. Cambridge University Press, Cambridge.

Lindsay, D.R. 1996. A Guide to Scientific Writing. 2nd ed. Longman, Melbourne.

Locker, K.O. 2003. Business and Administrative Communication. 6th ed. Irwin McGraw-Hill, New York.

Matthews, J.R., Bowen, J.M. & Matthews, R.W. 2000. Successful Scientific Writing: a Step-by-Step Guide for the Biological and Medical Sciences. 2nd ed. Cambridge University Press, Cambridge.

McMillan, V.E. 2001. Writing Papers in the Biological Sciences. 3rd ed. Palgrave MacMillan, Basingstoke.

Meadows, A.J. 1998. Communicating Research. Academic Press, San Diego.

O'Connor, M. & Gretton, J. 1999. Writing Successfully in Science. Chapman & Hall, London.

Peat, J., Elliot, E., Baur, L. & Keena, V. 2002. Scientific Writing: Easy When You Know How. BMJ books, London.

Penrose, A.M. & Katz, S.B. 2003. Writing in the Sciences: Exploring Conventions of Scientific Discourse. Longman, Harlow.

Silyn-Roberts, H. 2000. Writing for Science and Engineering: Papers, Presentations and Reports. Butterworth-Heinemann, Oxford.

Stapleton, P., Youdeowel, A., Mukanyange, J. & van Houten, H. 1995. Scientific Writing for Agricultural Research Scientists: a Training Reference Manual. WARDA/CTA, Wageningen.

Strunk, W. & White, E.B. 2000. The Elements of Style. 4th ed. Allyn & Bacon, Needham Heights, MA.

Sullivan, R.L. 1996. Technical Presentation Workbook - Winning Strategies for Effective Public Speaking. ASME Press, New York.

Swales, J.M. & Feak, C.B. 1994. Academic Writing for Graduate Students: Essential Tasks and Skills: a Course for Non-native Speakers of English. University of Michigan Press, Ann Arbor.

Swales, J.M. & Feak, C.B. 2000. English in Today's Research World: A Writing Guide. University of Michigan Press, Ann Arbor.

Turabian, K.L. 1996. A Manual for Writers of Term Papers, Theses, and Dissertations. 6th ed. University Chicago Press, Chicago.

Yang, J.T. 1995. An Outline of Scientific Writing: for Researchers with English as a Foreign Language. World Scientific Publishing, Singapore.

Zinsser, W.K. 2001. On Writing Well: The Classic Guide to Writing Non-fiction. 6th ed. Quill Press, New York.

JOURNAL ARTICLES

Bailer, J.C. & Mosteller, F. 1988. Guidelines for statistical reporting in articles for medical journals. Ann. Intern. Med., 108:266-273.

Gopen, G.D. & Swan, J.A. 1990. The science of scientific writing. Am. Sci., 78:550-558.

WEB SITES (ALL ACCESSED OCTOBER 2003)

General search engines

The Internet Public Library (collection of search engines)
http://www.ipl.org/ref/websearching.html

Scirus (scientific information search engine)
http://www.scirus.com

Web Ferret (combined search engines)
www.ferretsoft.com

Scientific communication

Online tutorial on all aspects of scientific writing and presentation
http://www.taa.org.uk/Courses2/CourseContentCSS/
CSSCourseContent3.html

Writing

Online Writing Lab. Purdue University, USA
http://owl.english.purdue.edu

Instructions to Authors in the Health Sciences. Raymond H. Mulford Library,
Medical College of Ohio, USA
http://www.mco.edu/lib/instr/libinsta.html

Literature searching

Ingenta (free literature searching of academic and professional journals)
http://www.ingenta.com

Oral and Poster Presentation

British Society of Animal Science (BSAS). Presentation Guidelines (preparation
of scientific posters, slides and PowerPoint presentations)
http://www.bsas.org.uk

Radel, J. 1999. Effective presentations. University of Kansas Medical Center, USA
http://www.kumc.edu/SAH/OTEd/jradel/effective.html

Mandoli, D.F. 1996. How to make a great poster
http://www.aspb.org/education/poster.cfm

Fisher, B.A. & Zigmond, J. 1999. Attending professional meetings successfully
University of Pittsburgh,USA
http://www.edc.gsph.pitt.edu/survival

Clip-art

Microsoft Clip-Art Gallery (clip-art that can be downloaded by licensed Microsoft users)
http://office.microsoft.com/clipart/

Clip-art for licensed users of Corel products
http://www.designer.com/files/default.asp

Other clip-art collections:
http://www.clipart.com/
http://clip-art.easy-interactive.com/

INDEX